Scales

A Collector's Guide

Bill & Jan Berning

4880 Lower Valley Road, Atglen, PA 19310 USA

Dedication

This book is dedicated to our children, Celeste, Nathaniel, and Jonathan, who put up with our endless hours of using the family computer. Thanks for your patience, guys! You are all the greatest!

Published by Schiffer Publishing Ltd.
4880 Lower Valley Road
Atglen, PA 19310
Phone: (610) 593-1777; Fax: (610) 593-2002
E-mail: Schifferbk@aol.com
Please visit our web site catalog at **www.schifferbooks.com**

This book may be purchased from the publisher.
Include $3.95 for shipping.
Please try your bookstore first.
We are interested in hearing from authors
with book ideas on related subjects.
You may write for a free catalog.

In Europe, Schiffer books are distributed by
Bushwood Books
6 Marksbury Rd.
Kew Gardens
Surrey TW9 4JF England
Phone: 44 (0)181 392-8585; Fax: 44 (0)181 392-9876
E-mail: Bushwd@aol.com

Designed by Randy L Hensley
Type set in UnivrstyRoman Bd/Dutch801 Rm BT

ISBN: 0-7643-0778-9
Printed in China
1234

Acknowledgments

We have had the strong support of many friends both old and new in our endeavor to complete this book. We could never give any of them enough thanks or tell them in words how much they have helped us. We appreciate each of you more than you may ever know!

We wish to acknowledge with deepest gratitude the following people and groups for their support and assistance in bringing this work to fruition: The American Egg Board, Jack I. Bender, Rick Berning, Dick Bueschel, Dee Crandall, Dennis and Vicky Hartwig, Erik Hewitt, International Society of Antique Scale Collectors, Jerry Katz, Edward Konowitz, William P. Lindeyer, Jan Macho, Mac and Jenifer McGilvery, Red and Barbara Meade, Roland Pautz, Larry Pugsley, Ned Schaeffer, Barry Silverstein, Roger "Jackpot" Simale, Bob Stein, Ted Stein, Hal Stern, Jeff Storck, United States Department of Agriculture, Burns Watling, Joe M. Wiley, Charles Yarton, and many contributors who wish to remain anonymous. We also wish to thank, with our sincere apologies, those who have accidentally been omitted from our listing.

Contents

Introduction

This book covers many types of scales. Contained in its chapters are examples of postal, shop, egg, coin, scientific, person weighing, household, toy, and other miscellaneous types of scales. There is also a non-scale category, containing items that are not scales themselves, but are related to scales in one way or another.

Although there are many scales pictured between the covers of this volume, no one book could possibly capture all of the examples of scales ever made. Most scale manufacturers produced hundreds of different scales and there were manufacturers in almost every country around the globe. The size of the scale companies varied from the largest such as Fairbanks, Avery, Salter, Howe, and Chatillon, to the small unknown scale makers who worked in their own barn. While many collectors are most readily drawn to scales from the last three centuries, there are drawings of scales being used as far back as the times of ancient Egypt as well. In addition to the wide range of time periods from which scales originate, scales also come in varying sizes. One scale pictured in this book is a mere one inch square, and then some scales are so large that they would be impractical for anyone to collect.

The scales shown here are representative of over seventy collections: beginner collections as well as some that are quite large and extensive. Some collectors we came in contact with specialize in a particular type of scale, while others collect scales of every category depicted herein. Consequently, you will find scales in this book that appear in 'as-found' condition, while others appear to be fully restored. There are collectors who prefer to collect scales that have been restored to better than new condition, while some prefer scales be left as found, especially if the original paint is in fairly good condition.

Price ranges indicated in this book are in 1999 United States dollars. The retail price range quoted is for a scale as it is shown in its existing condition. In the event that a specific scale is missing a pan or pans, weights, or other parts, is broken, or not in working condition, it will be worth less than a complete, working, accurate scale. In the event that a scale is in much better than average condition, restored, or with original paint in good to excellent condition, it will be worth more. As a rule of thumb, a common scale missing any parts, even minor parts, has a greatly reduced value, usually 50% to 90%. On the other hand, your proverbial "once in a lifetime" find, extremely rare, highly desirable scale missing a few parts or repaired will have less of a price deduction.

Most of the prices in this book are from private sales and auctions. Prices were also gleaned from catalogs, dealers or appraisers, International Society of Antique Scale Collectors price guide, and actual prices collectors have paid at antique shops and flea markets.

Pricing such a broad range of scales is far from an exact science. This is due to the extreme variations of age, type of scale, condition, rarity, and collectibility. Where you buy a piece is also a determining factor in the price. You might get lucky and find a Tiffany letter scale at a flea market, yard sale, or thrift store for a few dollars. At a fine antique shop, the same scale would sell for the full retail price. Bought from a collector, the price might be somewhere in between.

We, the authors, are scale crazy. We surround ourselves with our scales. We have them in every room of the house, give them to each other as gifts, and work with them daily as a full time occupation. We also go to flea markets, antique malls, and secondhand shops in search of other old or antique scales for our collection. There is always another rare scale just waiting to be discovered in a nook or cranny somewhere and as always many bargains to be had in the most unexpected places. We hope that within these pages there is something for everyone to learn or discover about scales.

In assembling these photos, we learned so much, talked to so many, and discovered that there are still hundreds of scales that we have yet to see. We know that our research and knowledge of scales only scratches the surface of what is out there waiting to be discovered. We hope this book will inspire you to become interested enough to begin your own collection or continue in your search for your next scale.

Please feel free to give us any feedback, ask us any questions, add to our knowledge about scales or point out any mistakes that you may find.

Bill and Jan Berning
Berning Scales
135 West Main St.
Genoa, IL 60135
E-mail: IweighU@yahoo.com
Telephone: (815)784-3134

We hope you enjoy our book! WEIGH DAILY!!!

Chapter One
Postal Scales

Postal scales are probably the single most popular group of scales to many collectors. The oldest letter or parcel scales date back to the British Uniform Postal Act of 1840. This was the first time a letter could be mailed based on its weight and not on the number of sheets of paper it contained. The United States started sending mail by weight around 1863.

Almost every type of weighing principle imaginable has been used in the manufacture of postal scales. There are postal scales that drop weights or lift weights as well as postal balances, steelyards, and spring scales a plenty. They range from crude utilitarian contraptions of questionable accuracy, to exquisite desk pieces of the finest woods and most precious metals.

This especially beautiful English postal balance is 8.75" wide. The base of this rare beauty consists of a porcelain plaque surrounded by an engraved brass frame. $3000 - $3500

The letter and weight plates as well as the base of this uncommon English letter scale have malachite inserts. The oval shaped inserts are framed by engraved and filigree brass borders. The ornate pillar rises from the center of the 6.75" wide base. This scale also has its full set of original weights. $3100 - $3400

This lovely gold-washed postal balance is signed "Tiffany" and was probably made by Mordan. It has blue and white Battersea medallions, depicting fairies, inset into the letter and weight plates. The original weights are missing from it. $450 - $550

This wonderful English postal balance made by Howell James and Co. Regent St., London, is embellished by seven blue Wedgwood medallion inserts. The 8.75" x 5.75" rectangular base holds six matched weights and is made of birchwood. The brass plates, pillar, and trim are highly pierced and engraved. $2000 - $2500

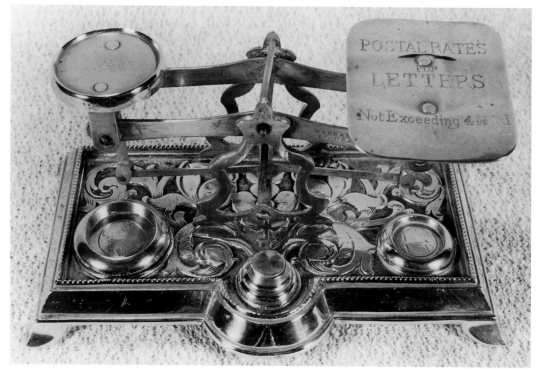

This brass English postal is marked "Warranted Accurate" and "Postal Rates for Letters Not Exceeding 4 oz." The engraved brass, rectangular base is 6.625" long and holds three weight receptacles containing a full set of five weights. $500 - $550

The only mark on this English postal is "Warranted Accurate." The brass scale sits on a rectangular base that is 8.75" x 5.375" and engraved in a floral design. It has three weight receptacles with one weight in each. $700 - $750

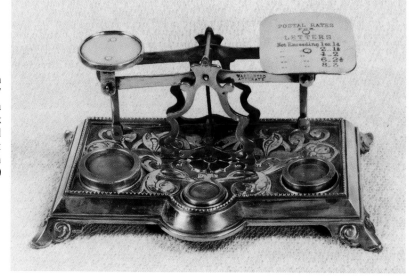

S. Mordan & Co. of London made this interesting postal balance. This simple scale sits on an oval black stone base under which there are 4 stone feet. The base has a mosaic overlay of marble and malachite. This brass scale's base is 8.5" wide and has an indented recess containing four matched nested weights. $900 - $1000

This highly engraved English balance scale sits on a 9.25" wide oval shaped base of black Italian marble. The base is inlaid with a lily of the valley design rendered in white and varying shades of green stone. $1100 - $1200

Shown here is a postal marked "S. Mordan & Co., London" and "Wilson & Gill, 139 Regent St., London." This silver English postal has a verde marble, rectangular shaped 11.125" wide base. The weight plate is recessed to fit the matched original weights. $975 - $1100

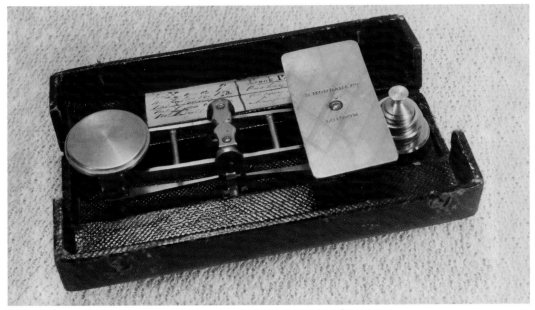

This brass English letter balance was made by S. Mordan & Co. of London. It is encased in a leather box with a hinged cover, and screwed in a fixed position to the bottom of the box. The oblong-shaped brass plate pivots to hold the four weights in place when the box is closed. There are hand-printed postal rates on the inside box cover. $675 - $725

Samuel Turner, Senior made this simple English letter scale, which also has an oval indention for weighing eggs. It sits on a legged wood base that is 7" x 2.875". The round pan is recessed to hold the four nested brass weights. $100 - $120

This English Air Weigh letter balance has a rounded silver-gray, cast metal base. It is 8.875" wide with a built-in receptacle at one end that holds five nested weights. The weights as well as the scale itself are brass. The scale has a crescent shaped label to determine equilibrium and postal rates. There are slots inside the base for inserting the scale and locking it in place. $80 - $100

This large English postal is brass and sits on a footed oak base. The parcel rates are inscribed on a plate affixed to the base. According to the postal rates, this scale was made from 1897 to 1906. $500 - $600

Pictured here is another English postal by S. Mordan & Co. of London. This scale is all brass and sits on a 15" long rectangular wood base. The letter plate is marked with postal rates and there is a complete set of original brass weights. This unusual scale has a hanging brass weight plate. $875 - $925

Shown here is an English, steelyard type, postal scale with a hollow rectangular brass beam. Its counterweight slides to reach equilibrium. It sits on a 10" wide pine base and is marked "S. Turner's Patent." $450 - $550

This silver plated 8-ounce letter scale has the name "Dan'l W. Tower" and "1885" inscribed on its beam. It stands 3" tall and 6.5" long. $350 - $375

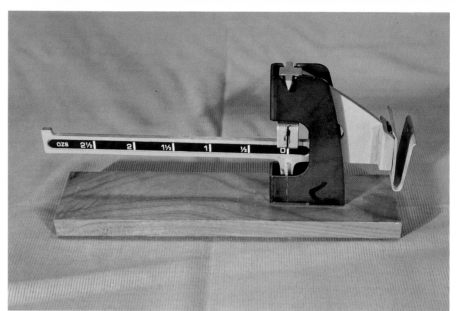

The maker of this letter scale is unknown. It is made of stamped steel and die cast metal. "The Santa Fe Magazine" appears on the 8" long wood base. $25 - $35

This mid-1800s postal has a 7" x 5" base of buhl inlaid with brass and tortoise shell. Believed to be of European origin, it has a kneeling cherub holding up a cylinder shaped beam. $3000 - $3400

Pictured here is an extremely rare English "ladder" scale made by DeGrave, Short, Fanner and Co., of London. It is made of brass and is set on a rectangular mahogany base that measures 16" x 5.875" deep. A ladder shaped frame holds three beams. The two lower beams have hanging plates that are fixed at 0.5 and 1 ounce weights respectively. The top beam with its hanging plate and pan is pre-loaded to 1.5 ozs. and has a total capacity of 9 ozs., achieved by adding the 0.5, 1, 2, and 4 oz. weights which are recessed in the base. $1000 - $1200

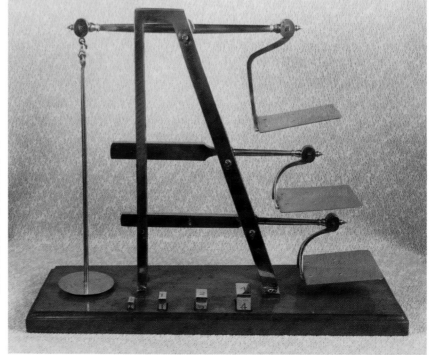

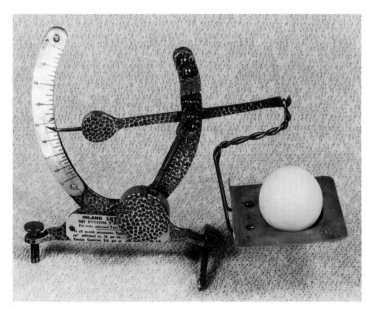

Samuel Turner, Sr. made this STS egg and letter balance. It has a celluloid plate with letter rates affixed to the base. It is 6.75" tall and has a peened, or hammered, effect brass body, weight, and indicator arm. The sheet metal graduation arch is calibrated from 0 to 4 ozs. The hanging letter tray has an oval concavity for additional use as an egg scale. $300 - $350

These three letter scales were made in Germany. They are quadrant scales from 5" to 8.25" in height and have a capacity of 2 or 4 ounces by one-sixteenth of an ounce. $60 - $70

This wooden based postal is marked "Deposé" on the weight arm and "E. W." on the v-shaped pillar. This hard-to-find French quadrant scale is graduated from 0 to 350 grams. $100 - $200

This rare brass letter scale is probably of German origin. It was most likely made for the British market since it has a 2 ounce capacity. It has a porcelain quadrant dial and was manufactured around 1910. $700 - $750

13

This French scale has a capacity of 60 grams. The artist's name on the bronze eagle is "A. Marionnet." The brass screw is used to level the scale. The height is 8.5" from the marble base to the top. $450 - $500

Day and Millward made this all brass postal with a wooden base. This rare scale uses a spring shock absorber so that the pendulum weight bounces when the load is removed from the letter plate. $250 - $300

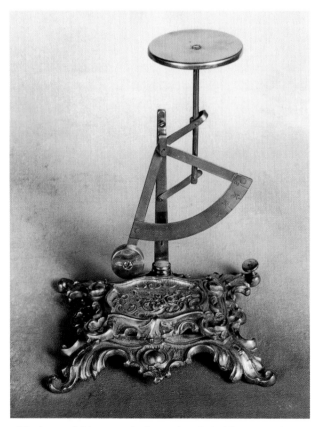

The base of this unmarked postal scale holds a stamp box. It is made of brass and has a 2 ounce capacity. $150 - $200

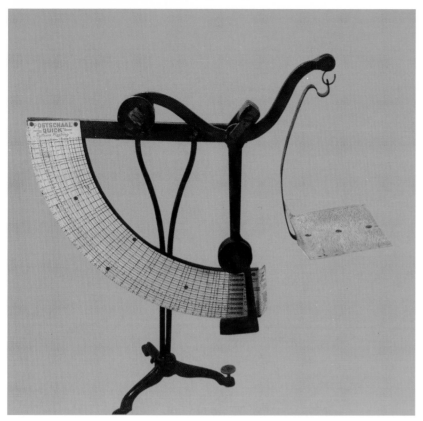

The Dutch postal scale is signed "Postschall Quick and System Keesing." It is made from cast iron with a white enameled dial and is calibrated in grams. It stands 21" high. $400 - $430

This hand made letter scale was probably made in France. The layout lines etched into the metal are still visible. It is graduated from 0 to 65 grams and is fairly accurate. It stands 5" tall. $40 - $50

Marked "DRP," this Austrian bilateral postal scale is made of brass with a black marble base. $200 - $225

This small German bilateral scale is marked "Made in Germany, U.S. Pat., Canada Pat. Mar. 20, 1906, Oct. 23, 1905, and Jul. 9, 1907" on the front. The back is marked "Columbus." It is nickel plated brass, graduated 0 to 0.75 oz. and 0.5 to 2.5 oz., and is 5.5" tall x 3" wide x 4" deep. It stands on bun feet. $40 - $50

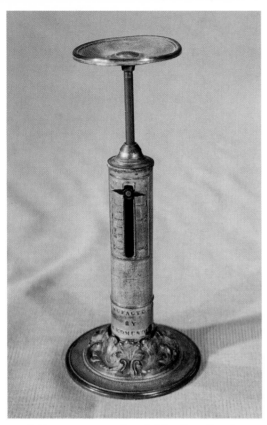

Collectors commonly call this a "candle-stick" scale because of the way it looks. This English made example is cast bronze with a capacity of 2 ounces. It was manufactured by Joseph & Edmund Ratcliff. $180 - $200

Shown here is an English spring balance signed "R. W. Winfield, Birmingham, Registered Sept. 5th, 1848." It is 7" tall with a brass mechanism and a body of Bristol glass with a gilded design. $350 - $375

This is an American spring scale manufactured by John Chatillon & Sons, New York. It was described in their 1894 catalog as a "combination letter balance and paper weight." This 5.5" tall cast metal scale has a decorative base and pillar. $140 - $160

This spring balance candlestick type scale is marked "Letter Balance Made in England" on its round brass letter plate. It stands 3.5" tall and is made of white metal with brass accents. It is 1.875" in diameter and is calibrated from 0.5 to 6 ounces. $50 - $60

This postal scale, manufactured by the Thompson Eng. & Mfg. Co., Chicago, Illinois, looks like an advertising paper weight, until you unscrew the top part, turn it over, and place it on top of the spring loaded, graduated post. It is made of aluminum and has a paper insert on the bottom with the current postal rates for first class, 3 cents per oz. In the closed position, it measures 2.75" high. It was made with a variety of different colored plastic bases: brown, black, and marbleized pink. The maximum capacity is 8 ounces. $35 - $45

Nolan Scale Co. of Boston patented this candlestick type letter scale in May of 1889. It is nickel plated with a circular dial that is 4.5" in diameter. The capacity is eight ounces by one-sixteenth of an ounce. $200 - $250

The Pelouze Scale Co. of Chicago made this sterling silver Countess Postal scale. It has a decorative design on the letter plate, base, and framing the face. A spring balance, it stands 2.75" tall and the base is 3.25" x 2". $325 - $350

This Victor Postal Scale was made by Pelouze of Chicago. It is made of thin, stamped, nickel plated, steel and has a capacity of 1.5 pounds. $25 - $35

This gold colored bronze scale is signed "Tiffany Studios, New York 872." It has a celluloid face, a capacity of 16 ounces and is 3.25" high. Shown in Tiffany's "Pine" design, it is made from etched metal over Favrile glass. The scale workings are marked and identical to the much less valuable Crescent scale made by Pelouze. $875 - $900

"Tiffany Studios, New York 1170," signed this 3.375" high postal. The celluloid face is marked "Crescent." It is made of bronze with Abalone shell inlays in the base and letter plate. This Tiffany pattern was named Abalone and the scale was once a part of a complete desk ensemble. $1100 - $1200

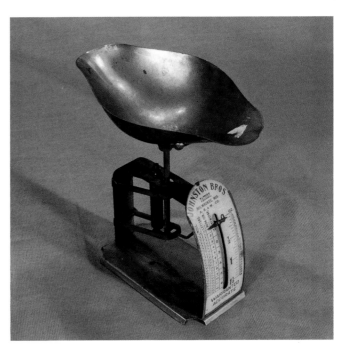

Tiffany Studios New York also signed this Star Postal Scale. It has a capacity of 16 ounces and shows a 2 cent first class letter rate. $850 - $950

The scale shown here was made by A. B. & M. Co. for the Johnston Bros. Factory in Milwaukee, Wis., and is "Warranted Accurate." It contains rates for merchandise, letters, books, circulars, and newspapers. The capacity of this spring postal scale is 1.5 pounds. $80 - $100

This ornate sterling silver postal has the hallmarks of Gorham. It is a half-egg shaped spring scale mounted on a wooden base. This 4" beauty has a "window" for viewing postal rates and bears the markings "Sterling B 2669 and Patented 1904." $600 - $700

Here is a small advertising letter scale for the Rexall stores. It is a small spring scale, 2.75" high, with a capacity of 8 ounces. The face shows 2 and 3 cents per ounce postal rates. Printed on the bottom is "Pat. Applied For." $15 - $20

This Hanson mail scale has postal rates for October 1, 1951. It is made of black and white plastic with an overall height of 6". $15 - $25

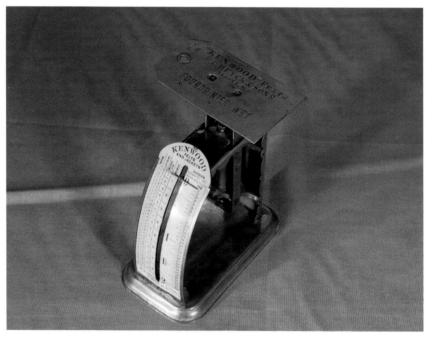

The letter plate of this postal scale contains an advertisement for Kenwood Felts, H. C. Huyck & Sons, Albany, NY, and is shaped like a shipping tag. It has a capacity of 2 pounds. $100 - $130

Shown here is a Bakelite desk organizer containing a 16 oz. letter scale, 2 compartments for paper clips or rubber bands, a pen holder, and a drawer for stamps. The name on the scale is Dennison, U.S.A., and it has a first class letter rate of two cents. $50 - $60

This Superior postal scale has a dust cover, which would come in handy in the office of a very dirty business such as a foundry or steel mill. The scale has a first class rate of 2 cents per ounce and a capacity of 4 pounds. It also has rates for merchandise, catalogs, books, and newspapers to different zones. It stands 7.75" high. The dust cover is very well made out of soldered tin and tole painted red with gold and white trim. It is 5.25" x 7.75" x 9.5" tall. $45 - $55

Here is a Reliance Postal Scale, "Patented Feb. 23, 1904, made by Triner Scale & Mfg. Co. of Chicago, U.S.A." It is a metal spring scale, tole painted black and silver, and is 7.125" tall. The 6.5" x 7" base houses two glass inkwells with metal covers, pen holders, and built-in drawers. $350 - $375

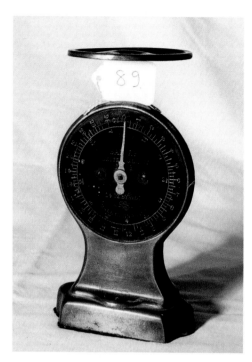

Salter made this English letter balance, number 41, for many years. It has a 3.25" letter plate and a capacity of 24 ozs. by 0.25 ozs. $75 - $85

D. R. Tompkins Manuf. of Pasadena, Calif., made this Stamprite Letter Scale. It is also marked "no Springs" and shows a 3 cents first class letter rate. This scale is different in that it lifts a weight for each ounce to reach its 8 ounce capacity. $35 - $50

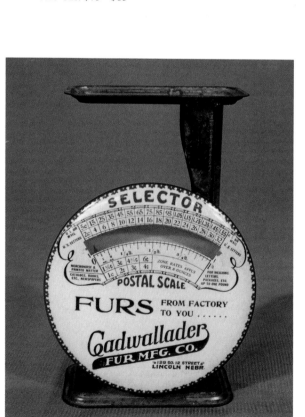

This Advertising postal scale is a very simple tin scale with no name or markings. It has a 2 cents per ounce letter rate, a capacity of 16 ounces, and is 5" tall. This one advertises Cadwallader Fur Mfg. Co. Many others exist. $80 - $100

The Sterling Adjustable Postal scale is made of plastic. It has a capacity of 11 ounces and shows a first class rate of 13 cents. $5 - $15

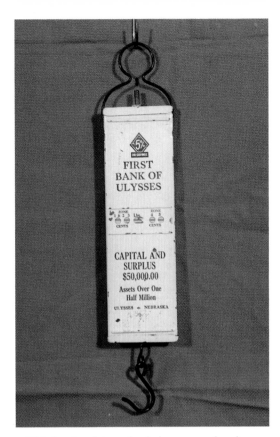

This unusual "Eldon Scale" is a hydraulic scale. It is filled with a colored liquid, which is visible in the front window. It has a capacity of 16 ounces and was made in the U.S.A. The 8 cents first class postage rate suggests this scale was made about 1972. $100 - $120

This is a tin plate spring balance parcel scale. This scale advertises the First Bank of Ulysses, Nebraska. Most of these scales carry advertising from businesses in Nebraska, Kansas, Missouri, or Iowa. $35 - $60

This clear plastic cube encases a 16 ounce letter scale. This is a very unusual scale bearing no makers name. $40 - $80

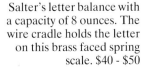

Salter's letter balance with a capacity of 8 ounces. The wire cradle holds the letter on this brass faced spring scale. $40 - $50

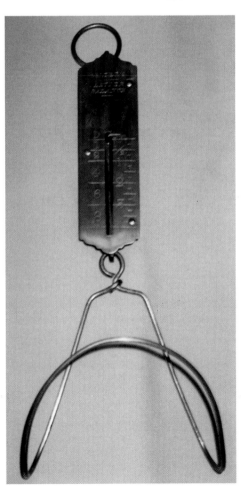

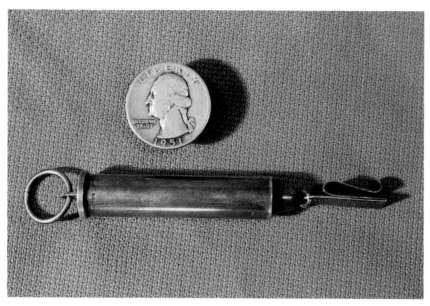

Tiffany and Co. made this small sterling silver postal.
It is 4" long and has a 3 ounce capacity. $150 - $200

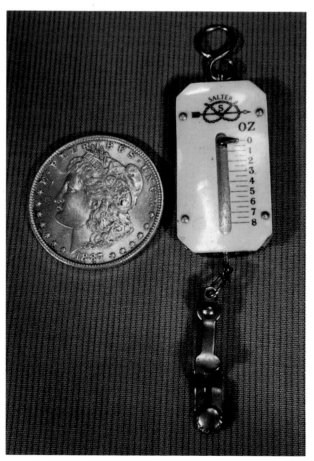

This tiny rare English postal has a
design registered in 1888 and was made
of nickel plated brass. An envelope
shaped door in the back opens to hold
stamps. It is 1" x 1.75" and the letter
clip is at the base. $300 - $350

Salter made this 8 ounce letter scale in England. This
small scale displays a clear example of the Salter logo, an
S inside a twisted rope with an arrow through it. $50 - $60

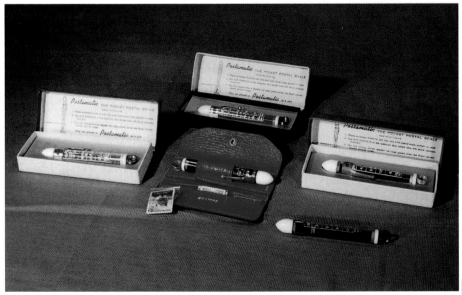

The Postamatic Postal Scale, plastic with a pocket clip and original box or genuine leather pocket case, was made in Phila., Pa., and is marked "Pat. Pend." It came in several colors - blue, yellow, clear, red, and green - and has a 4 ounce capacity. It appears to have been manufactured for several years because of the different first class rates: 2, 3, 4, and 5 cents. $8 - $12 each; $20 - $25 with case.

The Cross Pen Co. of Boston signed this quadrant, pendulum scale. It is made of nickel plated brass and has a 12 ounce capacity. $35 - $40

Pictured here is a handy, pen-like "Postalette." Made by the Exact Weight Scale Co. of Columbus, Ohio, it is 4.75" long and made of black plastic. When closed, it resembles a fountain pen whose top unscrews to reveal a letter clip fastened to the base of a spring rod. $55 - $65

This elegant sterling postal scale was made by Gorham from about 1891-1910. It has a capacity of 12 ounces and an overall length of about 6.5". It should have the original hook on top and also the letter clip. $850 - $950

Shown here is an Arc-Scale manufactured by the "Arc-Scale Manufacturing Comp'y, of Davenport, Iowa." This two finger held postal is two sided and has a flat, tin, wedge shaped body with a fixed corner weight. It has a hanging wire arm with a bottom hook and a paper label on each side with weighing instructions and calibrated postage costs. The patent on this hard-to-find postal is April 20, 1886. $175 - $200

C. H. and Co. of London made this "Universal Letter Scale." It is made of sheet brass and has a 6 ounce capacity. The scale shown here is missing the letter clip, which should be affixed to a small ring at the top of the teardrop shaped scale. $55 - $65

Thorne-Foster Inc. of New York City made this Sav-A-Stamp letter balance. It is nickel plated with an 8 ounce capacity. It measures 2" x 4.75" and shows a 3 cent first class letter rate. $25 - $35

This Sav-A-Stamp Postal Scale was made by A. Browne Co., 6242 Cherry Ave. in Long Beach, Calif. This flat scale is 6.375" long with a spring letter clip, pointer, and a finger ring for hanging the scale. The black metal face is calibrated from 0 to 4 ounces with postal rates indicating that first class postage is 3 cents per ounce. $20 - $30

Shown here is a hand held postal signed "Eschemann & Co., Phila., Pa., Pat. Oct. 3rd, 1899." It has a clip to hold the letter, is made of metal, and has a coated cardboard face advertising "Merchant's Metal Spanish Tiles." The total length of this advertising postal is 6.25", including its pendulum weight. $275 - $325

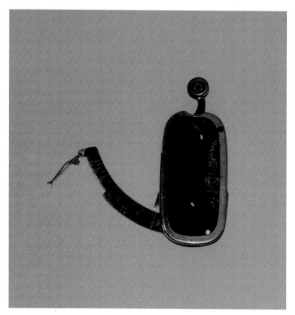

This little postal scale is marked "Focco" and "DRP." The complete scale folds into the oblong case enabling the user to carry it anywhere in his pocket. It is held by the round disk at the top and the letter is clipped into the letter clip. The case itself acts as the counterweight and the entire scale is only about 3" long. $30 - $40

This is a "Presto Letter Scale, Mfd By Metal Spec. Mfg. Co., Chicago, 24, Ill., Pat. Appl'd. for." This information is stamped on the bottom of this steel scale. The first class letter rate is 3 cents per ounce and the capacity is 5 ounces. From an engineering standpoint, this is a poorly designed scale because the letter being weighed rests against the side of the scale, causing the reading to be inaccurate. $25 - $40

Here is a small 2 ounce capacity letter scale with a round container used to hold a coil of postage stamps. The base is 5" long. $5 - $15

This hand made letter scale is made from wood scraps, nails, clothes pins, part of a file, wire, scrap aluminum, and miscellaneous hardware items. It has five steel weights marked in Roman Numerals, 0.5 oz. to 4 ozs. It is 11.5" tall. $30 - $40

Here is a bronze steelyard type English postal with a decorative round base and tapered pillar. The notched beam suspended from the pillar is held in balance by an L-shaped bracket. The counterweight hangs from one end of the beam and a U-shaped letter holder hangs at the other end. $350 - $400

The medallion on this very early English letter scale is imprinted "The Post Office Letter Weight" and on the obverse "For Rowland Hill's Plan of Penny Postage. Silvester & Co. 12 Aug. 1839." The overall height is almost 7" with a base diameter of 3.25". It has a letter clip attached to one end of the beam and a coin-like medallion that acts as the counterweight. $650 - $700

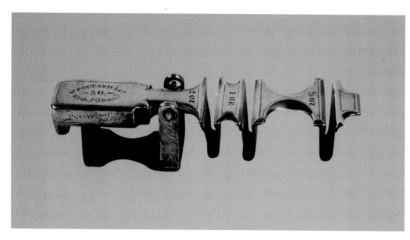

"W. Poupard & Co., 30 Wych St., Strand." made this English postal scale. It is 6.875" long, all brass with a stationary weight, and has three slots for letters, which each are calibrated: 0.5 oz., 1 oz., and 2 oz. $400 - $425

Shown here is a Mailway letter opener, which is also a simple letter balance. The letter opener acts as a counter balance to the slot into which the letter is placed for weighing. If the beam doesn't tip, the letter weighs less than one ounce. These scales were chrome, silver, or copper plated, and some were given or sold for advertising. $15 - $40

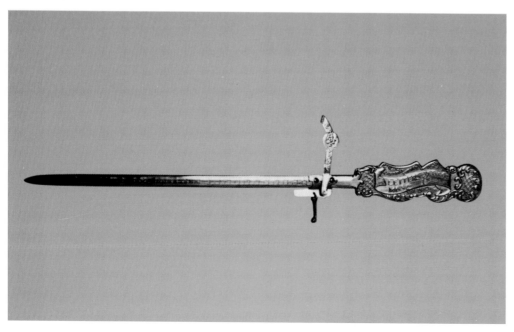

A Read's Postal Balance can also be used as a letter opener. It is made of silver metal and is 8.5" long. The decorative handle acts as a counter-weight attached to the beam that is calibrated from 0 to 20 oz. The metal slide attached to the beam has a suspended hanger arm and hanging letter hook, which move to locked positions on the beam when item is not in use as a scale. The marking on the handle is "Canada, Europe, U.S.A. Letters 2 cents Per Oz. Mdse. 1 cent Per Oz." $90 - $110

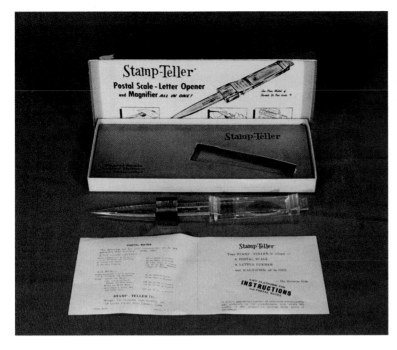

This Stamp-Teller postal scale is molded of DuPont Lucite®. The letter is placed on the handle and the metal poise slides along the blade to indicate the weight of the letter. It was made in New Canaan, Connecticut, in 1960. $20 - $30

Here is a hanging spring balance made by Triner Scale & Mfg. Co., Chicago, Ill., U.S.A. This scale is made from black metal, and has a chrome rimmed, glass covered, circular face. With a capacity of 20 pounds, it was used to weigh parcels. $120 - $140

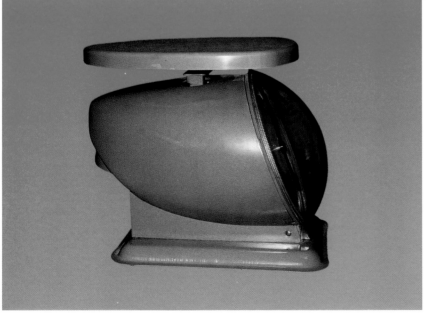

This is a side view of an unusual headlight shaped postal scale. It is an American Family scale with a 5 pound capacity and a first class letter rate of 5 cents per ounce. $45 - $55

Pelouze made this 2 pound capacity computing postal scale. It has October, 1953, rates for computing letters, air mail, parcels, and newspapers. $25 - $40

This Fairbanks 12.5 ounce capacity letter scale was made to commemorate the United States Centennial in 1876. It is made of cast iron and nickel plated brass with a very colorful painted design. It stands 4.5" tall. $150 - $200

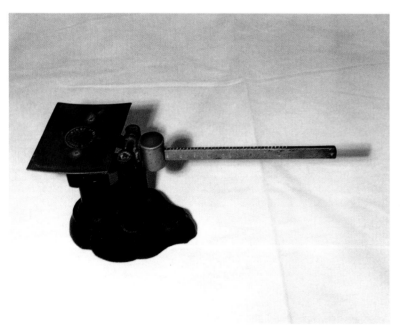

Imprinted on the letter plate of this 10 ounce letter scale is "Manufactured by B. Chambers Jr., Washington D.C." It is very crudely made of cast iron and brass, yet appears to have been mass produced. Some of the hand stamped numbers in the brass beam are in the wrong place, upside down, or completely missing. It stands 4.5" tall and the overall length is 10". $125 - $150

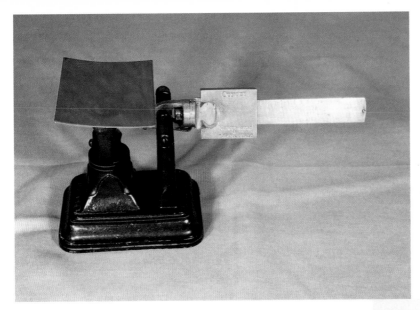

This Fairbanks Postage Indicator Scale is equipped with a revolving brass beam with postage rates for newspapers on one side and letters, books, and printed matter on the other. It stands about 4.5" tall. The 1906 Fairbanks catalog lists this 8 ounce capacity scale for $4.00. $125 - $140

The American Mach. Co., Philadelphia., made this Perfection Postal Scale patented Sept. 1, 1885. This 8 ounce capacity scale is mechanically the same as the 56 ounce version. It is made of cast iron with a brass letter plate on top. The small brass indicator can't be seen in this photo. $400 - $450

Here is an American Machine Co., Philade., Perfection Postal Scale, Pat. Sept. 1. 1885. This scale is patent number 325,534, issued to David Hallock. It has a very unusual mechanical design. There are eight weights suspended inside the base. For each ounce placed on the plate, one weight is lowered and this is indicated by the pointer on the right. It has a capacity of 56 ounces. It is 6.5" tall with an overall length of 15.75". The beam and poise are brass while the rest of the scale is cast iron, and painted black with red and gold trim. $300 - $350

This very popular Fairbanks double beam postal scale was made for many years to weigh letters or parcels up to 20 pounds. $175 - $200

Howe made this 64 ounce scale with a J. C. Bittschofsky lightning postage indicator attached to the brass beam. The poise (movable weight) has a wire attached to it to indicate the correct postage, in cents, on the large chart. It shows first, second, third, and fourth class mail rates on the front. On the back side, it shows the foreign rates. The base and mechanical parts are cast iron, while the beam, poise, and pan are brass. It is 8" tall and the overall length is 19". $150 - $175

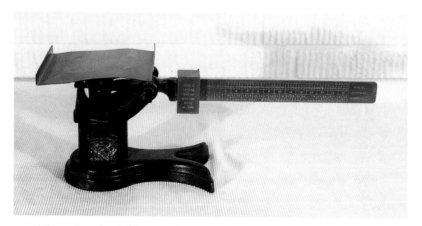

Shown here is a Weiss postal scale. The scale computes the weight of letters, newspapers, books, circulars, and merchandise, each on a different scale. It also states that foreign letter rates are 1 cent for 20 oz., and newspaper rates are 5 cents for 34 ounces. $100 - $125

The Computing Scale Co. made this brass and cast iron scale for postal use. The rotating chart indicates rates for letters, newspapers, and magazines. The beam on this scale is graduated to 2 pounds by 0.5 ounces and is missing its counterweight. $600 - $700

Shown here is a Fairbanks Express Package Scale. This small platform scale, 10.5" x 12", has a capacity of 55 pounds. In the 1906 Fairbanks catalog, this scale sold for $15.00. $120 - $150

Here is an equal arm balance scale marked "Canada Post Office." Its round bronze pan measures 4" in diameter and its square pan measures 5" x 5". $80 - $100

Triner Scale and Mfg. Co. made this model 805 postal computing scale. It is 15" tall, has a platform measuring 8" by 6.5", and a 5 pound capacity. It shows rates for first, second, and third class as well as air mail and foreign. $25 - $40

Chapter Two
Shop Scales

Shop scales incorporate those items that can be used for a wide range of things. They include the more common scales found in grocery stores in years passed, some of which may still remain in the produce areas of many large and small grocery stores. The small confectionery scales and the large barrel top grocer's scales are included here. Some of these scales are very desirable, while others require large spaces or are so heavy that collectors may find them unappealing or cumbersome to collect.

This Fairbanks confectioners' scale is quite unusual in that it has an inverted v-shaped beam on which the weight rides. The beam, weight, and pan are brass, while the scale is cast iron. Its capacity is 2 pounds by 0.5 ounces. It has a full capacity making counterweights unnecessary. In 1911, this scale sold for $8.00. $80 - $120

Fairbanks made this small confectioners' scale. It has a 1 pound capacity and is graduated by 0.5 pounds. It is made of cast iron with a brass pan and beam and sold in the 1911 Fairbanks catalog for $4.50. $70 - $110

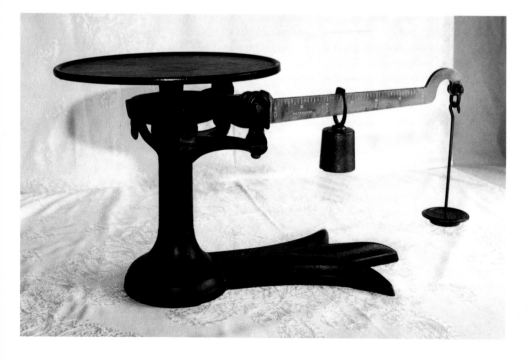

Here is a Fairbanks grocers' scale that was sold by Fairbanks for many years. It could be bought with or without a scoop and had a capacity of 31 pounds by 0.5 ounces. With the addition of separate counterweights, its capacity can be increased to 63 pounds by 0.5 ounces. In the 1911 Fairbanks catalog, this scale sold for $10.75, as shown here without the scoop. $60 - $90

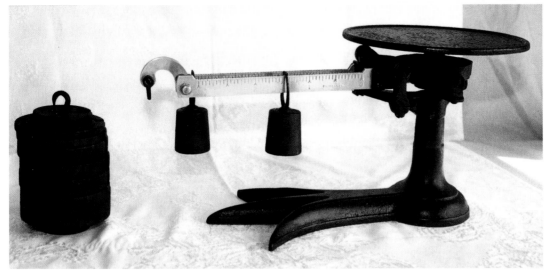

E. & T. Fairbanks made this double beam grocer's scale. Each beam of this scale had a capacity of 3 pounds by 0.5 ounces. With the addition of the counterweights, shown here, its capacity can be increased to 63 pounds by 0.5 ounces. This scale was originally sold with a tin, brass, or Russia iron scoop. $80 - $110

This Fairbanks counter scale came, according to the 1918 catalog, "equipped with brass or tin scoop, well supported upon a broad fork, so as not to tip easily." Its capacity with additional counterweights is 8 pounds, with the beam capacity of 5 pounds by 0.0625 ounces. It sold for $12.00 in 1918. $70 - $100

Jones of Binghampton, New York, made this interesting cast iron scale in the shape of a three leaf clover. The beam and counterpoise of this scale are brass. The cloverleaf weights are marked 6, 6, 6, 3, 2, and 1. This scale is 7" tall and was originally painted green. $120 - $140

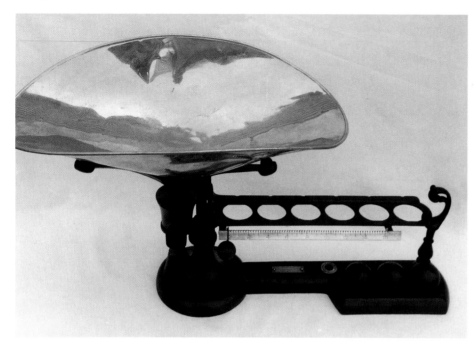

In the 1926 Troemner catalog, this Ball scale was described as a "Grocers' or general storekeeper's scale." The balls are used as counterweights, and its overall capacity is 16 pounds. It was patented on January 9, 1877, is 17.5" long, and made of cast iron and brass with a nickel plated pan. $150 - $250

The American Cutlery Company manufactured this "Sunbeam" cast iron scale with a diamond shaped base. It has a four pound capacity and was patented in 1906. $80 - $100

The only marking on this Canadian made scale is the name "Warren" on the base. The capacity is 10 pounds. The green paint with gold trim appears to be original. $100 - $125

This scale is marked "FERRETERA" and "LAPALMA." It is made of cast iron painted gold with green and red pin stripes. The beam, counterweight, and scoop pan are brass. The scale has a capacity of 1 kilo and is 8.5" tall, without its pan. $80 - $95

Manufactured by Henry Troemner, Philadelphia, Pa., this "Charles of the Ritz" powder scale is only 8" x 4". It was made to weigh loose powder in makeup departments of retail stores. The inspection sticker on this scale is from 1963 through 1966. It is graduated by sixteenths of an ounce to a capacity of 2 ounces. The number on the pan and scale must match. $120 - $140

This hand made scale is 7.5" tall and is 19" in length. It is made almost entirely of wood with the beam hand carved. The capacity of this scale is 6 pounds and the 0 adjustment was made by placing rusty square nails in a compartment under the tin pan. $100 - $110

The only name on this unusual scale is "Carters Standard." As the poise moves along the beam, the graduated brass face rotates to indicate the weight. The capacity is 18 pounds. It was originally red with black and gold trim. The height of the scale without the pan is 8.5". $400 - $450

This Howe Union, cast iron scale has a capacity of 44 pounds. It has its original red paint with gold trim and a brass beam. $90 - $100

Made by Fairbanks for many years, this Union Scale has two platforms. It originally came with a scoop on the smaller platform, which had a capacity of 30 lbs. by 0.5 oz. for finer weighings. The large platform, made for coarser weighings, had a capacity of 240 pounds by 0.25 pounds. The large platform measures 10.5" x 13.5". This scale was listed in the 1911 Fairbanks catalog for $14.50 with a tin scoop, or $15.50 with a brass scoop. $90 - $110

This is a rare platform scale made by John Weeks of Buffalo, N.Y. It is incomplete, broken, and rusty. The beam is graduated from 0 to 44 pounds on the lower scale and 0 to 5.5 pounds on the upper scale. This type of scale has little value except as a parts scale. $5 - $15

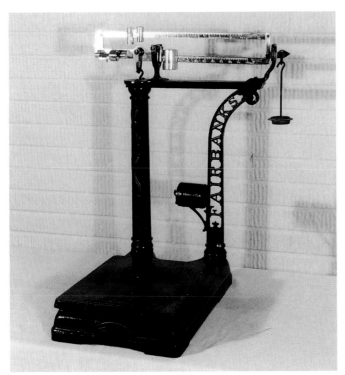

The Fairbanks Imperial grocer's scale shown here has a scoop balance at the left end of the beam for instant adjustment when the scoop is removed or replaced. The upper beam has a capacity of 10 lbs. by 1 oz., while the lower beam's capacity is 50 lbs. by 0.5 lbs. When weights are added, the scale capacity is 250 lbs. by 1 oz. The platform is 12" x 15", and it was made in black or red in 1911. $150 - $200

The Computing Scale Co. of Dayton, Ohio, U.S.A., made this very complex computing scale. This scale is made of cast iron and brass, with a marble tray. $200 - $250

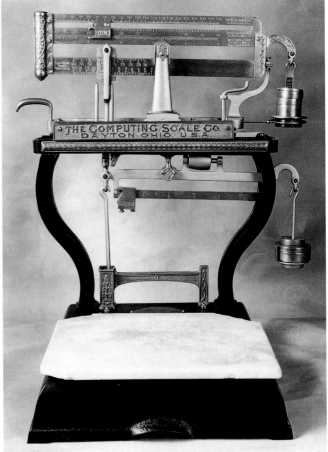

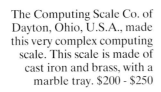

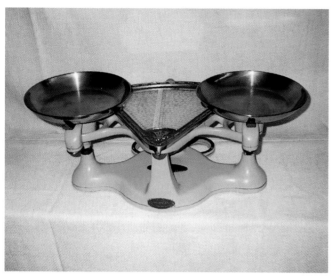

This unusual scale made by the Anderson Tool Company of Anderson, Indiana, is a computing scale. The price of the goods is figured on the chart by moving the poise to balance the scale. The patent date on this scale is October 2, 1906. This scale is very rare. $750 - $850

Shown here is a Premier Computing Scale, manufactured by the Standard Computing Scale Company of Detroit. The brass guide slides along the long chart to compute the price of the goods. It is hard to find one with all the weights. $325 - $375

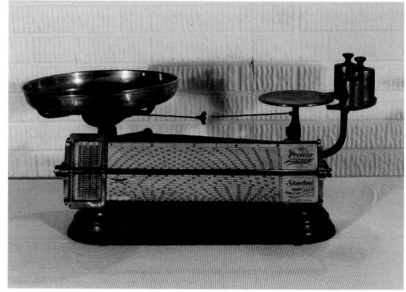

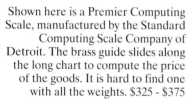

The Computing Scale Co. of Dayton, Ohio, made this hard to find scale. It sits on a brass base, which is capped by a marble slab. Made mostly of brass, it computes items costing from 5 cents to $1.00 per pound and has graduations to 2 pounds shown on its beam. It measures 13" x 7". $750 - $850

This interesting scale, which pivots on its base, was made by The Computing Scale Company of Dayton, Ohio. It is made of cast iron and brass and computes prices of goods from 5 cents to one dollar per pound. It shows patent dates of 1897, 1898, and 1902. $750 - $850

Dayton Scale Co. made this rare computing scale that collectors call a "beehive" scale. This fully restored scale has a swivel base and a rotating computing chart. The beehive shaped weights have levers to control their use. $1800 - $2000

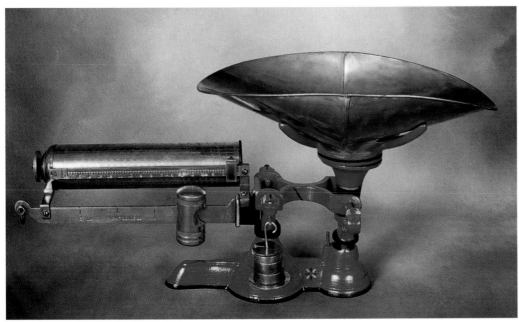

This cast iron and brass Acme computing scale is dated 1889. The brass cylinder can be turned to the correct position to compute the total price of the goods purchased. $500 - $600

National Scale Co. of Chicopee Falls, Mass., made this two pound capacity computing scale which computes prices from 5 cents to $1.00 a pound. The long barrel is rotated to the proper position to indicate the full price of the item. It stands 9.25" tall, with an overall length of 17.5". $400 - $450

The Standard Computing Scale Co., Ltd. of Detroit, Michigan, made this "barrel top" computing scale. This scale has its original gold paint and is made of cast iron with ornate brass details. It was patented in 1909. $150 - $300

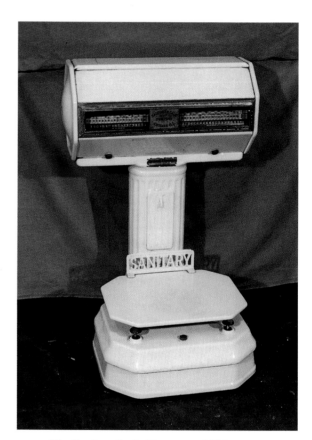

The Sanitary Scale Company of Chicago, Illinois, made this computing grocery scale. It is white porcelain on cast iron and has a capacity of 24 pounds. It is 31" tall and can compute the price of goods from 0.5 cents to 75 cents per pound. $150 - $200

Here is a large, 20 pound capacity Angledile scale with a closed face. These scales were also made with an open face, with the works visible to the public. Some of these scales were equipped with a fitted cast iron stand. $250 - $600

This Angledile Scale Co. scale is the rarest of the three sizes of scales that the company made. It has a five pound capacity and is a computing scale. It has a mirror to enable the customer as well as the shopkeeper to see the weight and price. This scale is missing the dial ring which is imperative in order to determine its value. The value quoted is for a complete scale. $5400 - $6000

Angledile Scale Co. of Elkhart, Indiana, made this scale to weigh 2 pounds. A confectioner's scale, it uses an oil filled dash pot for a damper. This scale, like others made by the same company, has a cone shaped rear dial that shows computations for cost of goods per pound. $200 - $400

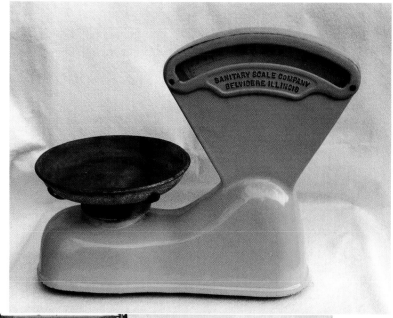

Commonly known to collectors as the "Boot Scale," this computing scale was manufactured by the Sanitary Scale Company, Belvidere, Illinois. It has a capacity of five pounds and is made of green porcelain over cast iron. Any repair work on this scale would require disassembly of the entire scale due to the one piece cast housing, which is why the scale did not last long. It was also available in white. It is 15.5" tall and 17" long at the base. $200 - $250

Toledo Scale Co. made this "fan" computing scale (with an overhead mirror). It is equipped with its original wooden base and has a 2 pound capacity. $250 - $350

The Brantford Scale Co. Limited of Brantford, Canada, made this computing fan scale. It has a 3 pound capacity and is 25" long. $150 - $250

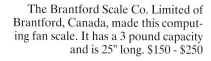

Detroit Scale Company made this restored confectionery scale. The fan shaped head on this 2 pound capacity, 22" long scale is slightly offset giving it an unusual appearance. $400 - $600

Stimpson made this offset base computing scale. This candy scale has a 3 pound capacity and computes from 8 cents to $1.25 per pound. It is made of cast iron and porcelain coated. $75 - $125

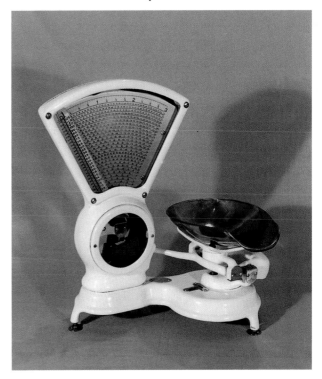

This Toledo Scale Company computing scale has been restored to better than new condition. It has a capacity of three pounds and computes the price of goods up to $1.00 a pound. $600 - $800; unrestored $100 - $150

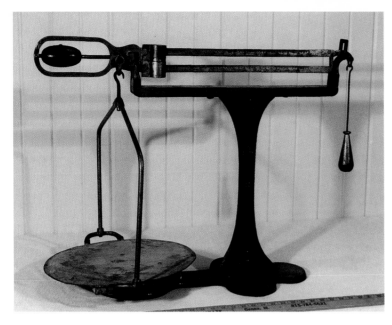

Buffalo Scale Co. made this double beam "Boston Market" scale. It was listed in their 1894 catalog and named "Best market scale-Gold Medal New Orleans 1884-85, Gold Medal World's Fair Chicago 1893." Its top beam is graduated 0 to 15 lbs., while the lower beam is graduated from 15 to 30 lbs. It was painted vermilion and had a 15" diameter marble slab plate. $75 - $200

The only name on this brass scale is "Lacave, Madrid." $250 - $400

This chain balance was made in Germany. The large dial can be turned to let out the chain to add more weight to one side of the scale. $200 - $300

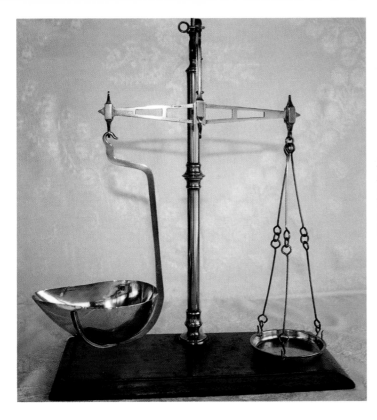

John White & Son, Auchtermuchty, Scotland, made this all brass scale. It is a class B tobacco scale and weighs 8 ounces. $200 - $400

W & T Avery Ltd. of Birmingham, England, made this "harp top" shop scale. It is made of cast iron and brass and has a ceramic goods plate. It has a capacity of 16 pounds. $250 - $500

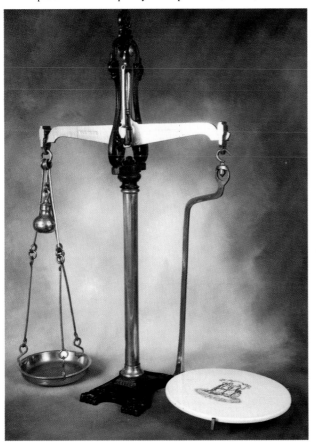

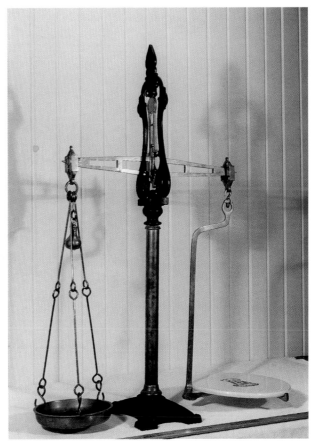

Avery of Birmingham, England, made this "harp top" market scale for the "Home & Colonial Stores LTD., London." Made of brass and cast iron, it is 40" tall and has a 22" long beam. This scale has a capacity of 15 pounds. $250 - $500

Testut of Paris made this hard-to-find French baker's or bread scale. It is made of cast, forged, and sheet iron with a brass pointer and a wooden base. This scale is 22" tall and has 12" diameter pans. $350 - $500

IBM made this 10 pound capacity scale. It has a fan shaped dial and is 27.5" tall. $125 - $225

Howe Scale Company of Rutland, Vermont, made this balance scale. It has its original red paint with gold trim. Made of cast iron with a brass beam, it stands 5.25" tall. A Howe scale from Brandon, Vermont, would be older than this scale from Rutland. $125 - $250

This scale is a Fairbanks "Patent # 1." It is 20" long, has an 8 pound capacity, and cost $4.50 in the 1859 Fairbanks Scale catalog. $150 - $300

Southwark Scale Company in Philadelphia, Pennsylvania, patented this rare scale in 1885. It has a brass beam and a revolving weight tray that is built into the base. $900 - $1000

This equal arm balance scale was made by C. Parker, Meriden, Connecticut. With a patent date of May 8, 1875, this cast iron scale is one of the more ornate from that period. $130 - $200

Philadelphia Scoop and Scale Mfg. Co. made this Hatch's Even Balance scale. It is cast iron and has octagonal goods and weight plates. It originally came complete with a scoop pan and 8 weights graduated from 0.5 to 8 ounces. $50 - $110

This interesting cast iron scale has no name or markings on it. $125 - $175

This is a very heavy marble cased scale. It has no name or markings on it, and the exact purpose is unknown. The base size is 10.5" x 25". $250 - $350

Believed to be of Spanish origin, this scale is made of marble and has brass pans. The only markings on this scale are "Capicidad" and "kilo." $225 - $325

This German market scale is only marked "3 KG." It is made of cast iron and has a floral design with two angels. The back of the scale is the same design as the front. This one is missing its pans. $90 - $130

Marked "Juliusz Sperling," this cast iron scale has a griffin design. The true weight indicators are shaped like sea horses, and the pans are brass. The large pan has a pouring spout on one side. $110 - $195

This interesting market scale was bought in Mexico and is marked "Sol" and "3." It is made of cast iron and has a 3 kilogram capacity. $95 - $140

The chief engineer for F. W. Woolworth Co. designed this scale and commissioned Jacobs Brothers to manufacture it. It was specifically designed for the bulk candy counter. To keep everything sanitary, it was nickel-silver plated. $350 - $450

Salter made this "Trade Parcels Balance Number 58T." It is made of cast iron with a 9" diameter brass dial graduated to weigh 100 pounds by one-half pound. This 17" tall scale was priced at £72 in the 1920 Salter catalog. $150 - $225

This scale is believed to have been made in China. There was much handwork utilized in the manufactory of this iron and brass scale. $45 - $60

This Imperial computing scale was made by Pelouze Scale Co. in Chicago. Its capacity is 2 pounds by 0.5 ounces. This scale was made for candy and tobacco and the Pelouze catalog states that the "dial computes candy or other merchandise in money from 5 cents to 60 cents per pound." The scoop measures 5.5" x 10" and the scale measures 4.5" x 7" x 10.5". $40 - $60

John Chatillon & Sons made this double faced circular spring balance with a 45" tall cast iron stand. In the 1900 Chatillon catalog, the stand sells for $2.50, while the scale is listed for $6.00. The pan was extra. $250 - $300

The Stimpson Computing Scale Co., Detroit, Michigan, made this ornate scale for computing the price of groceries. It is made of nickel plated brass and has a steel housing. It is equipped with a magnifying glass to read all the small numbers. $600 - $800

Shown here is a Stimpson computing scale. This type of scale with a vertical barrel is very rare. It computes the price of your groceries from 5 cents to 30 cents per pound. $700 - $900

The Dodge Scale Co. made this micrometer with a capacity of 5 pounds. All of the works are nickel plated. The patent dates are from 1892 to 1903. $350 - $450

This is a very unusual cast iron scale with no name or markings. The brass levers at the right flip weights on to the scale. The scale also incorporates a beam across the front with a brass poise. $500 - $600

Dodge Manufacturing Co. of Yonkers, New York, made this micrometer. This very accurate scale is made of cast iron with a brass pan and other details. It was patented Nov. 22, 1892; Mar. 22, 1898; and July 21, 1903, and has a 20 pound capacity. $300 - $400

The Perfection Scale Company of Philadelphia, Pennsylvania, made these Perfection scales in several sizes. Both of these examples have their original paint and were patented February 3, 1885. The smaller size, 2 with a 2 pound capacity, is more desirable than the larger size, 4 with a 20 pound capacity. $300 - $500

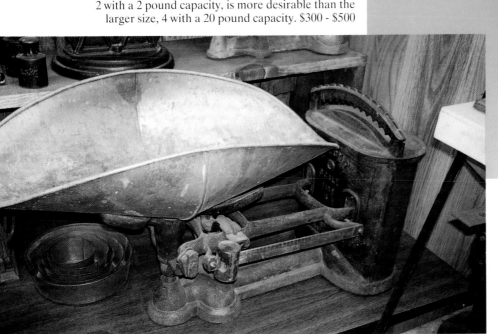

This "Turnbull's, Patent Sep. 13, 1859" has a double sided brass face. Below the face in gold paint is "To weigh 8 lb. by 0.5 ounces." The base of this hard-to-find scale is 13.5" long. $200 - $300

Here is a class C English Bread scale to weigh 4 pounds. It has fork like prongs with which to hold the bread. These scales were usually sold with a self containing metal or wooden box, so they could be stored neatly on a baker's cart. $175 - $250

Eggs are weighed primarily for two purposes. A vast majority are weighed for the purpose of placing them in the proper size category for commercial sale. Most fertile eggs produced are intended for hatching purposes. Because egg size is directly related to both hatchability and chick size, eggs are also weighed so that the most desirable may be incutated.

When actually packing eggs, it is usually necessary only to determine if individual eggs meet the minimum weight for the size category in question. This is why the "go, no-go" type of scale speeds up the weighing process, while scales that use a pointer to indicate individual egg weights are not as time-saving. This is also the reason many scales have a stop of some sort that limits scale action, effectively speeding up the weighing process. Experienced operators are also likely to become adept at placing individual eggs in the proper size class without actually weighing each and every egg. In short, when weighing hundreds or thousands of eggs per day, generally it is much more important to be able to place them in the proper size category than it is to weigh them with accuracy.

Historically, the terms "grade" and "grading" when applied to eggs have taken on a variety of meanings, and at times have been applied to any process that sorts eggs according to standards involving weight. Standard egg weights, as we know them today, came into common usage at both the state and national level during the 1930s and the 1940s and have remained essentially unchanged during the intervening years. Since eggs have traditionally been sold by the dozen, egg weights are most commonly stated in ounces per dozen and are actually size ranges.

Size	Weight Range
Peewee	15 oz. to less than 18 oz.
Small	18 oz. to less than 21 oz.
Medium	21 oz. to less than 24 oz.
Large	24 oz. to less than 27 oz.
Extra Large	27 oz. to less than 30 oz.
Jumbo	30 oz. or larger

In practical terms, egg packers think of these as minimum egg weights for a particular size.

Some scales also have graduations in points. A two ounce egg would be 100 or 1.00 points. These points are actually percentages with two ounces being equal to 100%. This system, which never did catch on, was used as a standard for pricing eggs, evaluating hen performance in egg laying contests, and as a formula for adjusting egg quantities in cooking recipes.

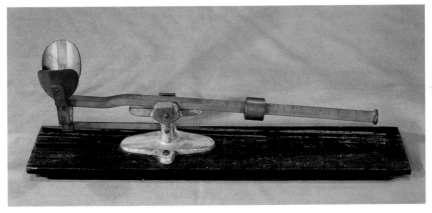

This Reliable egg grading beam scale has a capacity of 16 to 32 oz. The 12.75" x 2.5" wooden base is cut to fit on top of an egg crate. It has a cast iron stand with all other parts made of brass. In the 1936 catalog of Hoeft & Company of North Chicago, Illinois, this scale sold for $1.50. $110 - $160

The Royal Manufacturing Company, of Toledo, Ohio, made this one size hatchery scale to determine eggs 23 ounces per dozen and larger. This entire scale is made from stamped sheet metal and painted green. "23 ounces" is stamped into the base which is 3.75" in diameter. This scale was available in other sizes and was featured in Royal Manufacturing Company's 1929 catalog. The dealer's price at that time was $4.20 per dozen. $75 - $90

This scale has the following printed on its paper label: "Standard Egg Scale made by Petaluma Electric Incubator Co., Petaluma, California." It has a galvanized sheet metal beam and egg cup. A lead weight on the other end is marked "22" (ounces per dozen). The overall length is 8.75", and the un-painted base is 4.5" long. $75 - $95

Brower Mfg. Co., of Quincy, Ill., made this "Save All" egg scale. The entire scale is made from green and yellow painted sheet metal. The yellow painted beam is graduated from 18 to 30 ounces per dozen. Its sliding weight can be adjusted for a predetermined minimum egg weight. The base is 2.25" x 7" and has a 1.5" diameter hole at one end. $85 - $110

Stamped on the metal base of this scale is "MADE BY THE OAKES MFG. CO., TIPTON, IND." This entire scale is made of sheet metal with a brass bearing pin. The beam is graduated from 16 to 26 ounces per dozen with an adjustable sliding weight. There are many variations of this scale, including a "one size" 23 oz. version. This scale was offered in the 1938 catalog of the Brown Fence and Wire Co., Cleveland, Ohio, for 53 cents, postpaid. $30 - $45

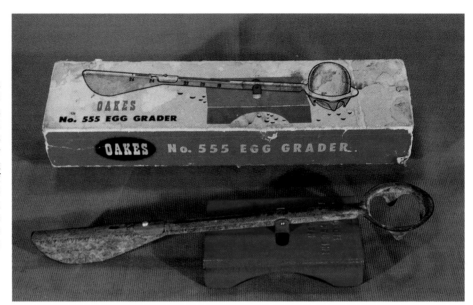

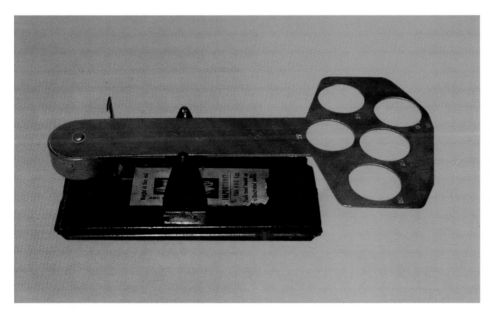

This A.B.C. egg scale is equipped with five holes in the beam, each marked with a different weight. It is made of thin sheet metal, mounted on a wooden base, and is probably of Canadian origin. $70 - $90

Pictured here is a Hart Tru-Way egg scale. The metal information plate has "Tested for 19-22" on it with the remainder of the line unreadable. Also included on the plate are complete instructions for operation. It was manufactured by H. W. Hart Mfg. Co. of Santa Monica, California. It is made from galvanized sheet metal strips attached to a green painted wood base. An auxiliary weight can be attached to the bottom of the fixed lead weight at the end of the beam. The addition of this weight increases the size range by one ounce per dozen. The length of the base is 10". $75 - $95

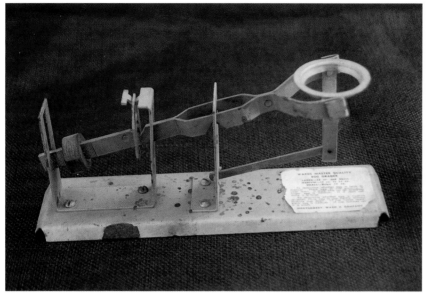

Montgomery Ward & Co. sold this "Wards Master Quality Egg Grader, Large-22 oz. and above, Medium-19 oz. to 20 oz., Small-below 19 oz." All of this information and complete operating instructions are printed on a paper label attached to the base. The general design and some of the parts of this scale are very similar to the Hart Tru-Way egg scale. The base on this scale is made of sheet metal and is 9.75" long. $70 - $95

The Reliable egg scale was made by the Reliable Mfg. Co. of Los Angeles. The cast metal beam and stand are attached to the stamped sheet metal base. This base is shaped to fit on top of an egg crate while the scale is in use. This scale uses various combinations of a flip weight, two lead weights, and a ring weight to weigh eggs from 18 to 30 ounces per dozen. This scale is tough to find with a complete set of weights. There is also another version on a wooden base. $55 - $85

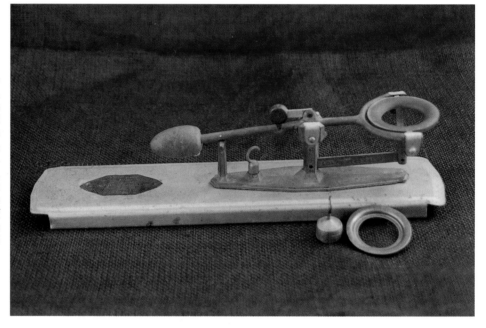

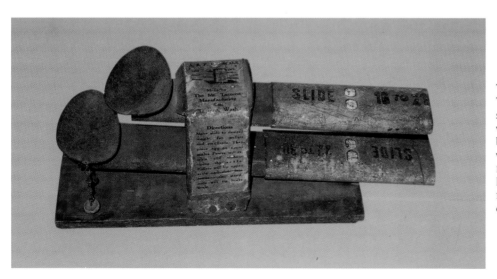

The A.B.F. egg scale of Mount Tacoma Mfg. Co. is made of thin sheet metal and mounted on a wooden base. Each of the two beams has a telescoping sliding weight. An egg that is too heavy for the top beam tips onto the lower beam. Complete directions for using this rare scale are printed on the side. $400 - $450

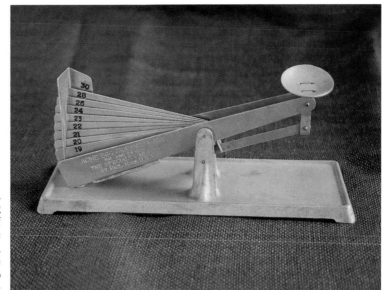

Specialty Mfg. Co. of St. Paul, Minn., U.S.A., made this "Acme Egg Grading Scale, Pat. June 24, 1924." Both of these Acme egg scales work on the same principal. The heavier the egg, the more thin aluminum strips are picked up on the other side. One strip is lifted for every graduation on the scale. The biggest difference between the two scales is one has the graduations on the individual weights, while the other has the graduations on the end of the weight guide. These scales are made almost entirely of thin aluminum and are very fragile. The Acme egg-grading scale was offered for sale at $3.00 in the February 1927 issue of the *Reliable Poultry Journal.* $20 - $35

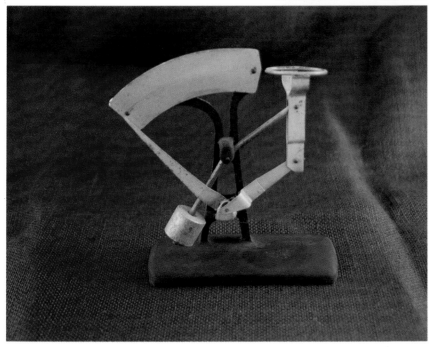

Prospectus Mfg., Minneapolis, Minn., made the "Mascot Egg Grader" shown here. All markings have been obliterated. Originally, it was graduated from 16 to 32 ounces per dozen. $40 - $50

The Oakes Mfg. Co. of Tipton, Ind., made this scale of unpainted galvanized sheet metal. There is an adjustable counterweight to calibrate the scale. It is graduated from 18 to 30 ounces per dozen and 1.5 to 2.5 ounces each. The brass screw in the base serves as an adjustable stop. This scale sold for $1.18, postpaid, in the 1938 catalog of the Brown Fence and Wire Co., Cleveland, Ohio. $45 - $65

Here is a "Mascot Egg Grading Scale, Pat. Appl." It has a blue painted base and frame of stamped sheet metal. It is graduated from 18 to 30 oz. per dozen and 1.25 to 2.5 oz. per egg. It includes an aluminum pan and pointer and has a cast metal beam with a fixed counterweight. This scale has a screw in the vertical slot that serves as an adjustable stop. It is 6.5" tall with a base of 6.25". A bent wire suspended in round cut-out in the scale frame serves as a leveling device. $45 - $65

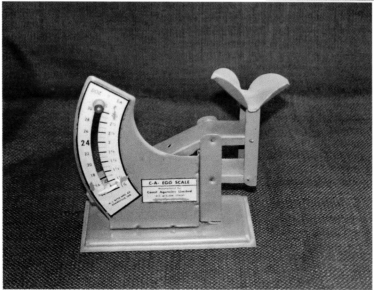

H. J. Otto Mfg. Co., of Evansville, Ind., made this egg scale. On its decal is printed "C-A- Egg Scale manufactured for Coast Agencies Limited, 1353 Willow Street" (city has been obliterated). The entire scale is made from stamped sheet metal and painted light brown. It is graduated from 15 to 30 ounces per dozen and 1.16 to 2.5 ounces each. This scale is very similar to one made by Brower. H. J. Otto also made this scale with the decal "Lilly's Grader, The Chas. H. Lilly Co., Seattle-Portland-Yakima." It was painted light green. $35 - $50

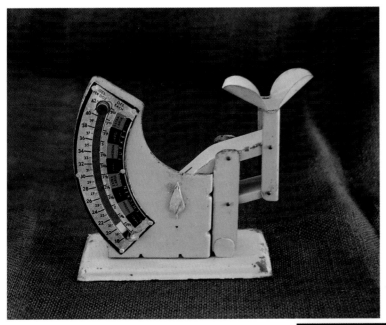

Brower Mfg. Co. of Quincy, Ill., made this "U.S.A. Brower's '3-IN-1' Egg Grader." The company claimed that it had the "World's largest line of poultry supplies." This is a dual purpose scale for weighing both chicken and turkey eggs. The aluminum arc is marked 18 to 42 ounces per dozen and 1.5 to 2.5 ounces each. An adjustable stop allows the scale to be set for a maximum weight to speed up weighing. It is marked for small, medium, large, and extra large chicken eggs as well as small, medium, large, and extra large turkey eggs. The base size is 6.5" x 3". In the 1949 Brower catalog, this scale was offered for $1.75, postpaid, in the U.S.A. $45 - $65

The Brower Mfg. Co., of Quincy, Illinois, produced this painted sheet metal scale. An egg placed on the pan raises the graduated aluminum arc to indicate the egg's weight. It is graduated from 18 to 30 ounces per dozen, 1.5 to 2.5 ounces per egg, and 70 to 130 points. A swivel stop allows the scale to be set for 22 or 24 ounces per dozen. A small steel ball in a cup covered with a plastic window serves as a leveling device. Its base size is 8.5" x 3.5". This scale has been seen in several different colors from different manufacturers or distributors. $45 - $65

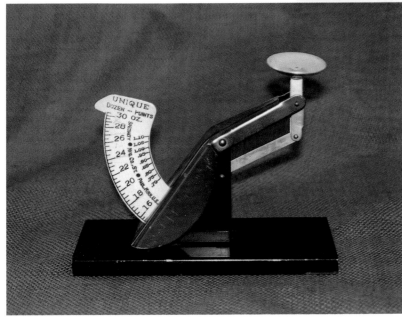

The scale shown here is marked "Unique Egg Scale, Specialty Mfg. Co,. St. Paul Minn. U.S.A." The entire scale is made from sheet metal and painted brown, except for the quadrant that is painted white with black numbers. It is graduated from 16 to 30 oz. per dozen and from 0.70 to 1.10 points. There were several variations and colors of this scale. $35 - $45

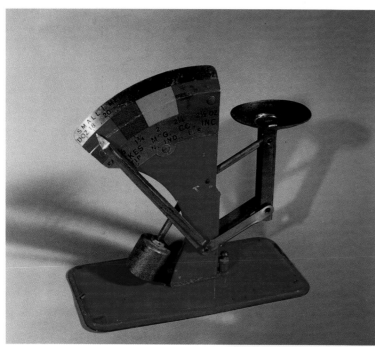

The Oakes Mfg. Co., Inc. of Tipton, Ind., U.S.A., made this scale. It is made of stamped steel and painted red. It is graduated from 17 to 30 ounces per dozen and 1.5 to 2.5 ounces each. The dial is color coded for the four different sizes of eggs. The brass screw in the 3" x 6.5" base serves as an adjustable stop. $40 - $65

Jiffy-Way Products, Inc., Minneapolis, Minn., made this scale, claiming to be the "World's largest mfg's of egg scales." It is made from stamped sheet metal and painted red or green. It is graduated from 18 to 30 ounces per dozen and 1.5 to 2.5 ounces each. The cotter pin suspended in the slot serves as a leveling device. It also has a screw that serves as an adjustable stop. This scale has patent no. 2205917, from 1940, and the base is 3" x 8". This is one of the most common egg scales and can be found with many different names—Farm Master, Wards, and DeKalb to name a few. This scale is also made in an almost identical modern version. The face on the old scale is a decal, while the new ones use a heavier sticker. $30 - $45

Prospectus Mfg. Co. of Minneapolis, Minn., made this "Mascot Egg Grader." It weighs eggs from 16 to 32 ozs. per dozen. All of the works in this scale are enclosed in a one piece cast aluminum housing, eight inches tall. With its soft round corners, this is one of the easiest egg scales to keep clean, a necessity when working with dozens or thousands of eggs at a time. $95 - $110

Here is an egg scale made by John Chatillon & Sons, New York. This dial type spring scale can be used to obtain quite accurate egg weights, since it is graduated from 0 to 4 ounces by 0.10 of an ounce. Extremely accurate weights are not required for most commercial purposes; therefore, this scale is probably intended to select hatching eggs or for poultry research which requires fairly precise egg weights. Overall height is 10.5 inches. $110 - $125

This Chatillon egg scale is one of the few American made spring egg scales. It has two graduated weight scales. One is identified as "weight per dozen in oz." (1.3 to 2.7 oz.). This scale is made from stamped sheet metal and painted black. Base is 5.5" in diameter and overall height is 8". $125 - $145

Salter made this "Spring Balance No. 11T, Made in England." It is made of green painted cast iron with a brass face that is painted white. The egg cup is brass and the scale has a capacity of 8 oz. by 1 gram. The gray circle on the 6 (at about 9:00) is a lead seal that the inspector would stamp. The overall height of this scale is 8.25". $110 - $125

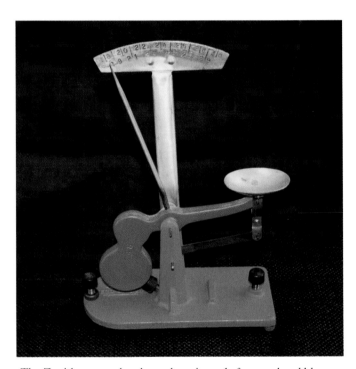

The Zenith egg grader shown here is made from red and blue cast metal. The scale has a leveling screw and two rubber stops that can be adjusted for rapid determination of egg weights. It is graduated from 18 to 30 ozs. per dozen. This scale was also manufactured in a freely suspended version. $75 - $95

This Salter spring scale, no. 18E, was made in England. It is graduated from 0 to 48 ounces per dozen and from 0 to 4 ounces each. It has a zero adjustment screw on the top, under the egg cup. This scale housing is similar to others made by Salter for other weighing purposes. $110 - $125

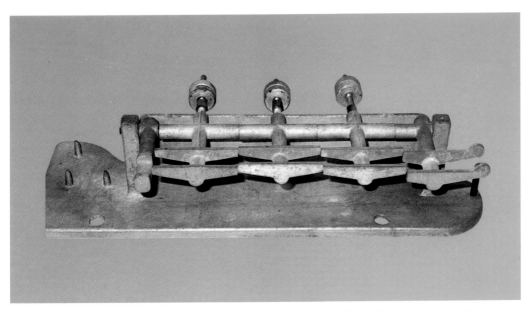

This automatic egg grader has no name or markings. The scale is used to separate eggs into three different weight categories. It is operated by rolling the egg down the two rails. Each of the three sections on the rails is a separate scale. As the egg rolls down the rails to the correct scale, that section tilts down to drop the egg into a receptable. $100 - $200

Shown here is a Jiffy-Way egg scale with candler. This is the usual Jiffy-Way scale, base and all, mounted on top of a candler. The flap on the lid in front of the scale reflects light from a long slit up to the face of the scale. $40 - $55

This is a Reliable Automatic Dial egg scale. Its brass arc is graduated from 16 to 40 ounces per dozen. It has a sheet metal body with a wooden base; both of which are painted green. The extended base allows the scale to be positioned across a standard wooden egg crate. The height is 8.75". The scale with the candler also has an adjustable stop. Described in the 1936 Hoeft and Company catalog, it sold for $3.00. Without the candler, it sold for $1.70. $100 - $150

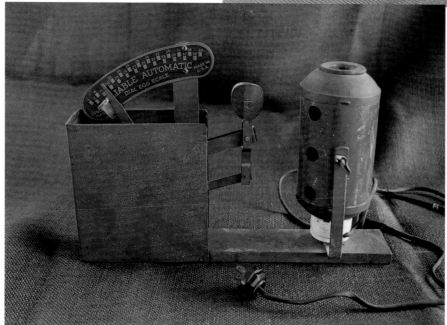

Gradencandle Mfg. Co. of Lynn, Mass., made this poultryman's light weight sheet metal scale called a "Gradencandler." The capacity is 18 to 30 ozs. per dozen. The addition of a light bulb adds another feature to this scale, a candler. The size of the base is 8" x 4.5". $110 - $135

Chapter Four
Grain Scales

There are many types of grain and seed testing scales; all of which are commonly referred to as grain scales. The main task of the grain tester is to calculate the net weight of the grain per bushel. This weight is sometimes referred to as a Winchester bushel, which is a unit of dry measure equivalent to 2150.4 cubic inches. The standard for this measure was originally deposited in the English city of Winchester, from whence its name originates.

A grain moisture balance is used to determine the percentage of moisture removed from grain. A set amount of grain is weighed, dried, and then reweighed. The scale user then calculates the percentage of moisture contained in the original sample. A percentage scale is used to calculate the percentage of dirt, hulls, leaves, insects, etc. in a sample of grain. The required amount of seed is placed in the scale container, weighed, and then sifted to remove the impurities.

The cleaned seed is then placed back in the scale container and the poise moved to a balance. The percentage of contaminants may then be ascertained accurately from the graduations on the beam. These scales could also be used to calculate the percentage of moisture. Additional names for them are corn or bean scales, cornometers, or chondrometers; these vary depending on the manufacturer and geographical location.

This U. S. Standard scale weighs 70 lbs. x 1 lb. and 2.25 lbs. x 8 ozs. This scale is marked "Winchester Bushel," which is a unit of dry measure equivalent to 2150.4 cubic inches. It has no connection to the famous gun maker. $150 - $250

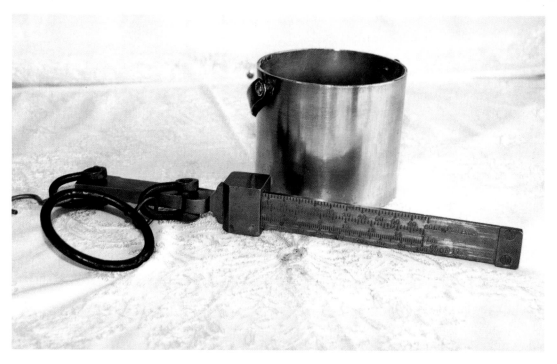

The Buffalo Scale Co. of Buffalo, N.Y., made this grain tester. It has three scales on the beam: 0 to 65 lbs. per bushel, 0 to 1 lb., and 0 to 100% of 1 lb. This scale sold for $15.00 in the 1928 and 1929 Buffalo catalogs. $250 - $375

This grain scale has both Howe and Ohaus marks on it. It measures 2 quarts in its brass bucket. It is graduated from 0 to 65 pounds per bushel, 0 to 4 pounds by one-quarter pounds, and 0 to 100%. Its weight is based on the Winchester bushel. $250 - $375

This Fairbanks seed tester is made of brass and cast iron. It is used to determine the "pounds per bushel and percentage of purity of beans, peas, seed or grain of any kind." The brass container measures one bushel based on one gill cup measure. The three scales are graduated: 1) lower - 100 x 1%; 2) middle - 4 oz. x 1/4 oz. for ordinary weighing; 3) upper - 64 lbs. x 1 lb. This scale sold for $8.00 in the 1911 Fairbanks catalog. $275 - $400

F. A. Thiele of Kopenhaven (Copenhagen) made this Danish grain tester. For storage, the three brass cylinders (*left*) fit one inside the other and can be stored in the mahogany box. The beam and weights (*right*) fit into the smaller box, which stores inside the larger one. $250 - $500

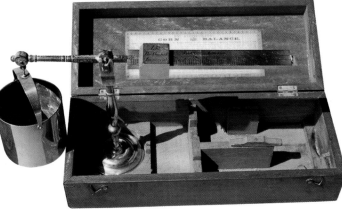

Loftus of London made this small corn balance. It weighs the pounds per imperial bushel of corn based on a small sample. The mahogany box is especially fitted for the scale and has a label inside the hinged lid giving instructions for use as well as mean weights for many different grains. $500 - $600

Lewis Schopper made this boxed grain moisture tester. For storage, the tubes fit together with the beams inside; both can then be placed into the drawer. This scale was made in the 1880s. $500 - $600

This Troemner grain scale fits into a wooden box especially furnished for it. Made mostly of brass, this rare scale attaches to a stand which is fitted on the top of the box. $600 - $800

H. Kohlbusch of New York made this chondrometer which is marked "Winchester" and came in its own fitted wooden box. It mounts on the box during use. This scale was not made by the famous arm's dealer; rather, it is based on a unit of measure and the English city of Winchester where the standard was originally deposited. $750 - $850

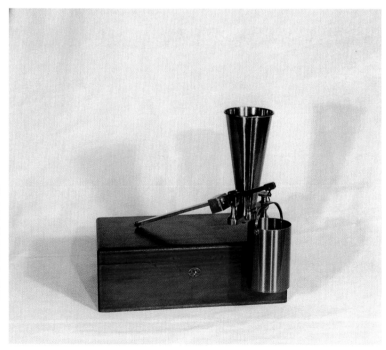

Dairy, milk, and cream scales all perform the same function. These scales are made especially for use in connection with the Babcock test for keeping records of the quantity of milk from individual cows. Most of these scales are equipped with two dial hands. The red hand can be adjusted anywhere on the dial to offset the tare weight of the empty pail. The majority of dairy scale capacities are 30, 60, 90, or 120 pounds. To make mathematical calculations easier, most dials show one pound graduations and are broken down by half-pounds and tenths of a pound rather than ounces.

This plain brass faced milk scale has brass plates riveted to it to become an advertising scale for Purina Cow Chow. It has a capacity of 30 pounds. $110 - $125

Chatillon made this "Improved Circular Spring Balance Milk Scale." Its brass face is 4.5" x 10", and it has a capacity of 30 pounds. $35 - $45

The Purina Cow Chow advertising milk scale shown here is adorned with the familiar Purina checkerboard pattern. The length of this scale, including hooks, is 15.5", and its capacity is 30 pounds. The design is embossed on the brass face and painted black. $125 - $155

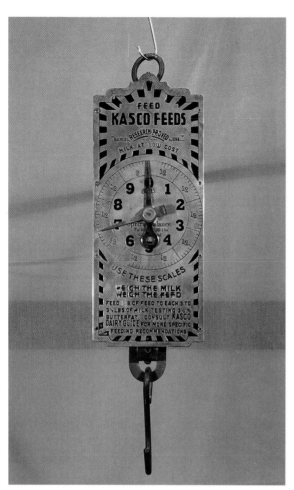

Shown here is a Kasco Feeds advertising dairy scale. This decorative brass faced spring scale has a capacity of 30 pounds. In order to get the net weight of feed or milk, the red hand would be preset to the tare weight of the empty container before weighing the full container. $120 - $150

Shown here is the "Sandringham Dairy Herd Recorder" by the Dairy Supply Company Ltd., Museum Street, London. This large brass faced dairy scale is 10" in diameter. It has a capacity of 41 pounds, 32 pints, or 4 gallons. The Royal dairy is located in Sandringham. $95 - $110

This Globe Milk-Producer advertising dairy scale is labeled "Globe Milk Meter" on its brass face. It has a capacity of 30 pounds. The gray circle at the upper left is a lead seal for an inspection stamp. $120 - $150

Hanson Scale Company, Chicago, made this Purina advertising dairy scale. It is a spring scale made out of thin steel with a capacity of 60 pounds. It is 8" in diameter. $45 - $65

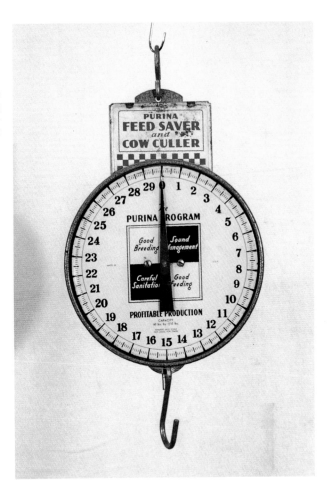

This Pillsbury's Best Feeds advertising dairy scale is made out of thin steel. It is 7" in diameter and has a capacity of 60 pounds. $45 - $65

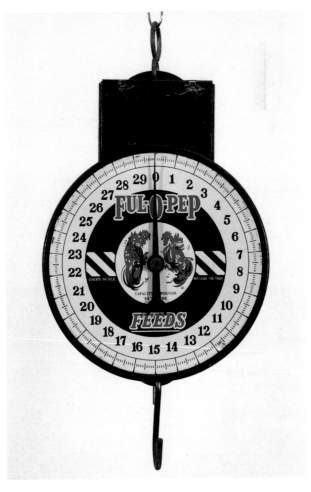

Shown here is a Ful-O-Pep Feeds advertising dairy scale. This colorful scale is probably made by the Hanson Scale Company of Chicago. The dial hand makes two revolutions of the 30 pound chart for a capacity of 60 pounds. It is made of painted tin and depicts two fighting cocks. $75 - $85

Coin scales were necessary to determine if coins were real or counterfeit in the days when coins were made of gold and silver. Some coin weighing scales determine the dollar amount of a large number of coins of the same denomination. These scales were a must-have for the vendor collecting from his route of coin operated machines.

Bank scales were useful for weighing gold or silver bullion. Some were also used to weigh bulk coins. Miners would bring gold or silver to banks or assay offices to be weighed in order to determine the cash value.

Diamonds and other gemstones were weighed using gem scales to determine their carat and to assess their worth.

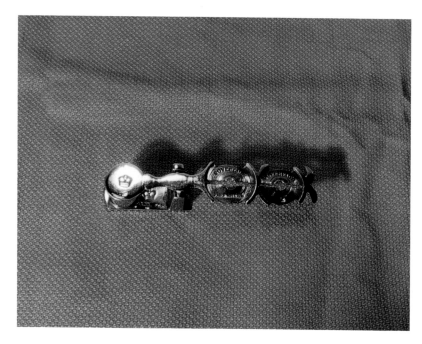

Hallmarked with a crown, this counterfeit coin detecting rocker is made of brass. The round spaces were used to check the exact diameter of sovereign and half sovereign, while the slots were to check the thickness of the coin. It is about 3.5" long. $85 - $125

This counterfeit coin detector was made by C. E. Staples, Worcester, Mass., and patented by H. Maranville, Clinton, O., Jan 13, 1857. This small scale is used to weigh U. S. and foreign gold and silver coins. Concentric rings were used to check the diameter of the coin and the notches on the edge were used to check the thickness. This all brass scale is 4.5" long when extended to the U. S. silver dollar point. $250 - $350

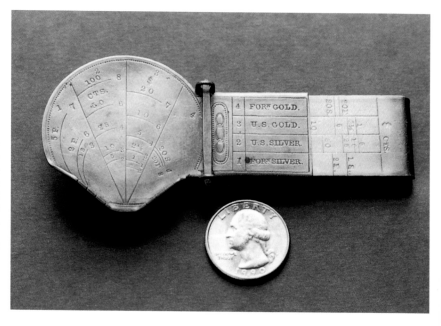

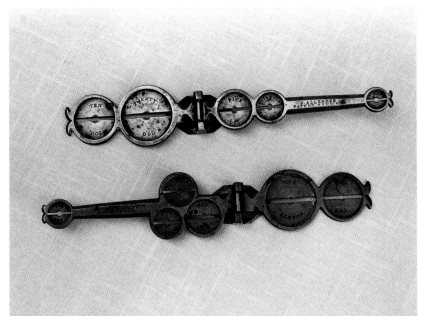

The five coin brass rocker balance designed by John Allender is marked patent pending. It is made for detecting counterfeit American gold coins. There are separate diameter and thickness gauges for gold $20, $10, $5, $2.50, and $1 coins. The six slot counterfeit coin detector on the bottom has a three dollar diameter and thickness gauge added to the one shown above. It was patented by John Allender on November 27, 1855. $250 - $375; six slot $300 - $425

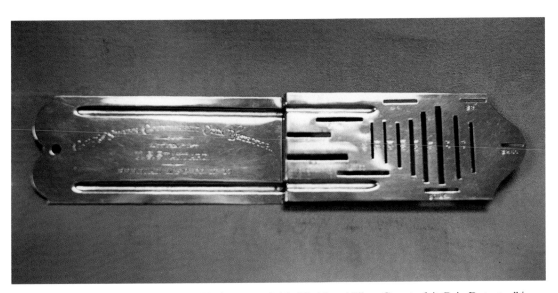

Berrian Manufacturing Co. of New York patented this "Gold and Sliver Counterfeit Coin Detector" in 1877. The slots in this simple scale are used to check for correct size of United States coins. $300 - $400

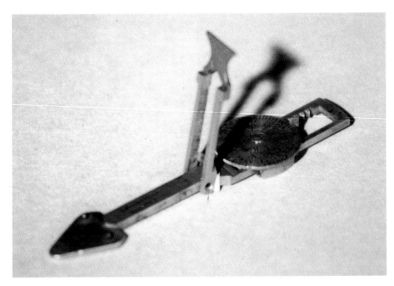

This rocker scale is shown in its open position. Most of this type of scale were made in the Middle East and were used to check coins for proper weight. When the scale is in use, the handle extends upward, as shown, and the round dial can be turned to the proper coin denomination. $150 - $250

Shown here is a cruciform rocker used for detecting counterfeit coins. Scales of this kind were made for use in the Middle Eastern countries of Turkey, Greece, and Egypt. It has two beams that are hinged together; one beam used to hold the scale, while the other is used for weighing. $90 - $110

A. Wilkinson made this folding sovereign scale. There is a label inside the box explaining its use. This scale has two different settings determined by a flip over weight hinged onto the top, left hand side of the scale. $300 - $400

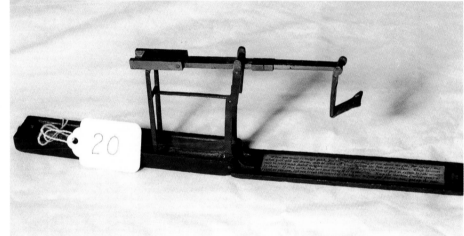

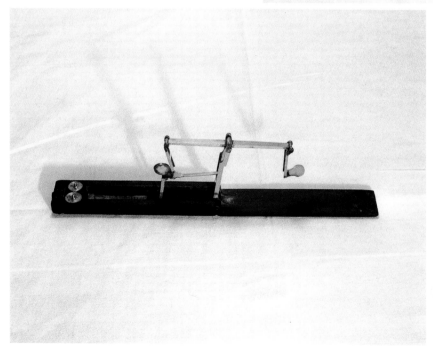

Wilkinson of England also made this folding gold balance, which can be folded into the wooden box to which it is affixed. The box also has two recessed compartments to hold the weights used for determining a guinea and half-guinea. $300 - $400

This banker's scale was made in the seventeenth century. The original box contains a complete set of 24 weights and has a label affixed inside its lid. $1200 - $1500

This German coin scale with a complete set of weights stores neatly into this wooden box. It was made before 1850. $600 - $700

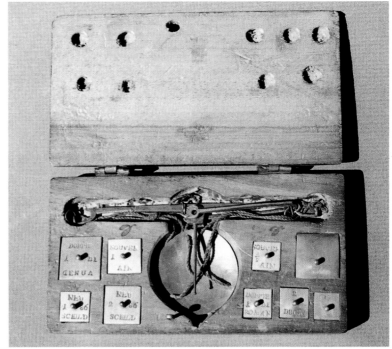

Salter Scale Co. of England made this Silver and Copper Checker, no. 53. This spring scale was used to compute the total value of pennies or silver coins. It is 12" tall and has a capacity of £100 in silver or £5 in pennies. $75 - $150

Hanson Scale Co. made this Penny and Nickel Counter for vending route operators to compute the value of pennies and nickels. This model 1538 has a capacity of $7.50 in pennies and $23.65 in nickels. The nickel value graduations are shown in red on the scale's face. $25 - $35

Shown here is a Hanson penny counter scale. The scale, pan, and accessories all fit inside the carrying case. It has a capacity of over $14.00. $25 - $50

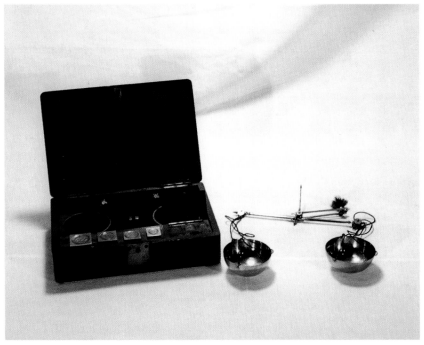

Degrave, Short & Co., London, made this diamond scale which was distributed in the U. S. by J. Dickinson, Nassau St., N.Y. It has a fitted mahogany box to hold its weights which are graduated 1 to 500 carats. $150 - $300

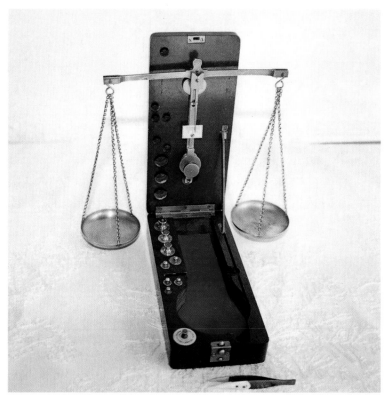

The traveling jeweler's scale shown here can be disassembled and fits into the 9" long wooden box along with the weights. "William Dixon, Inc., N.Y., Daube & Hopken" appears printed on the cover over the small weights. $150 - $200

W. T. Avery of Birmingham, England, made this gold scale with a steel beam and brass pans and weights. The scale, pans, and weights all fit in this oak box with the maker's label affixed to the lid. $130 - $150

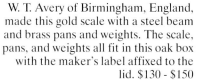

There is no identification on this very small diamond scale. The entire scale and set of weights fit in the 2.5" long wooden box. $150 - $250

This is a "miner's improved gold scale manufactured expressely for California." The small brass scale and its weights fit inside the tin box which has a gold painted eagle on its cover. $180 - 200

The United States Mint made this fully restored scale. It has a capacity of 2000 ounces, and the scale itself weighs about 1600 pounds. $ unknown

Shown here is a one-of-a-kind, all brass scale built by the Philadelphia Mint in 1917. It was on display in the assay office for many years. $ unknown

Chapter Seven
Scientific Scales

Scientific scales were used to weigh chemicals in a laboratory, dentist's or doctor's office, or for a scientific experiment. Most of these scales were, by necessity, very accurate. They include, among others, precision balances, torsion balances, and analytical balances.

Shown here is a Christian Becker analytical balance. This type of scale was made during a wide range of years; many are from 1910 to 1930. $275 - $350

Henry Troemner made this analytical balance, which might also be referred to as a prescription scale or balance. All legal terms, its names varied from state to state over the years. The scale is very accurate and was used by pharmacists, botanists, and other scientists. $450 - $550

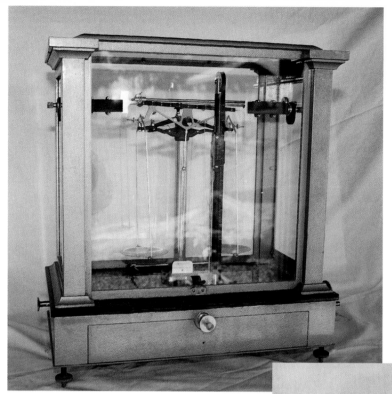

W. J. Ainsworth & Sons, Inc., of Denver, Colorado, made this type BB, precision analytical balance. $250 - $325

This Galileo scientific scale was made in Milano, Italy. This is a very complex and accurate scale. $250 - $350

Henry Troemner made this lovely wood and beveled glass torsion balance. This extremely accurate scale has a matched set of brass pans. It measures 20" long x 10" high x 10" wide. $295 - $325

Listed as selling for $21.00 by Troemner in their 1913 catalog, this scale has a cabinet made of marbleized slate. The art of marbleizing slate was done only in Fair Haven, Vermont, from the late 19th to the very early 20th century, making this scale hard to find. Even without the 8" diameter pans, it is a fine example of a box scale. It has a capacity of 10 pounds. $150 - $250

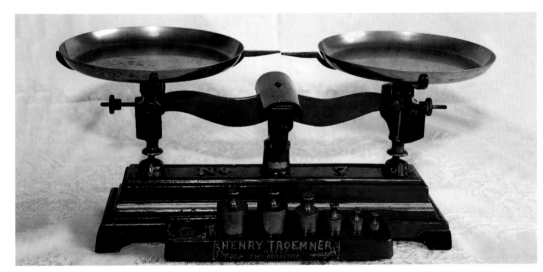

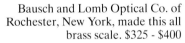

Henry Troemner of Philadelphia made this pharmaceutical and laboratory scale. It is equipped with moveable nickel pans, will carry 1 lb. in each pan, and is sensitive to 1 grain. There is a "full set of weights from 1 grain to 8 troy ounces, neatly fitted in a projecting shelf attached to the base." It is 12.5" x 3.75" and has a capacity of 12 ounces. This scale sold for $10.00 in Troemner's 1913 catalog. $150 - $200

Bausch and Lomb Optical Co. of Rochester, New York, made this all brass scale. $325 - $400

Ohaus made this triple beam laboratory balance. It is cast iron with brass beams. $50 - $75

The Torsion Balance Company of New York made this Torsion Balance. The brass beam has a capacity of 10 pounds, and the scale is balanced by thin steel bands. Patent dates on this scale are from 1885 to 1891. $90 - $110

This Torsion Balance lab scale has a single pan with a beam by which to determine the weight of the item being weighed. $40 - $60

Fairbanks made this Harvard trip scale. This type of scale was originally designed for Harvard University for use in physics and chemistry lab experiments. It is fitted with two porcelain plates and a brass side beam with a capacity of 5 grams, graduated in one-tenth grams. The total capacity of the scale is 1 kilo. The price in the 1911 Fairbanks catalog, without weights, was $7.50. The name painted on the front of the scale is Knott, Boston. The height is about 6". $85 - $125

Pelouze Manufacturing Co. of Chicago, IL., U.S.A., made this "No. 8 Rexo Scale." It is sensitive to one grain and has scales for avoirdupois and metric weights. $35 - $50

This Fairbanks prescription scale was patented April 23, 1878. With its pan, it has a capacity of 8 drachmas or 30 grains (pan is missing in the picture). $95 - 110

N. V. Randolph & Co. made this small scale and its spatula, which fold into the nickel plated brass box for easy transportation in one's pocket or a doctor's bag. Stamped into the top of this box is "Dr. Fitch's Prescription Scale, Pat'd Sept 29, 1885." With a capacity of 20 grains, this scale would have been used by doctors or pharmacists for dispensing medicine. The box measures 3" x 1.5" x 0.75". Other versions of this scale omit the manufacturer's name from the lid. $145 - $165

Kerr Dental Manufacturing Company made this dental scale with its own stand. The scale in the photo is missing its shallow pans. $25 - $50

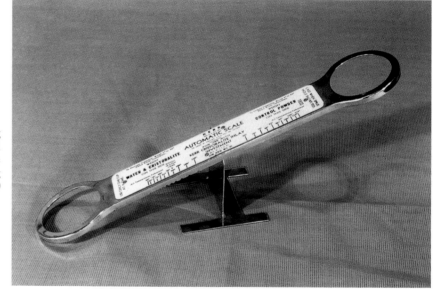

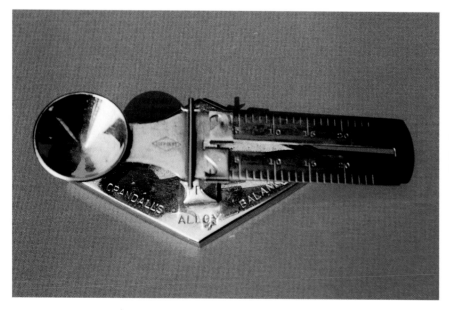

This Crandall's alloy balance was used as a dentist's scale. It has one beam with two scales, each with a poise. $125 - $145

This Hanson, model no. 1475, was made in Northbrook, Illinois. This is a white plastic diet scale with a bowl and its own carrying case. The scale has an adjusting nut on the top and a capacity of 200 grams or 7 ounces. The plate lifts off and fits in the case along with the bowl and scale. The scale measures 6.5" tall, and the case is 5.75" tall. $35 - $50

Pelouze made this dietetic laboratory scale, which measures 8" x 3.75" x 6.75". Its porcelain face is graduated 4 ways: 1) Metric - 300 grams by 2.5 grams; 2) Apothecary - 9 oz. by grams; 3) Avoirdupois - 10 oz. by 0.125 oz.; and 4) Avoirdupois - 10 oz. by drams. The scale comes with its own fitted case. $40 - $55

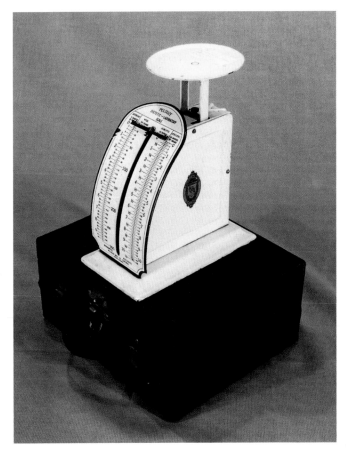

Chapter Eight

Person Weighing Scales

Person scales were made to weigh babies, patients, people on the street, and even people in their home bathrooms.

Coin operated scales also known as penny scales or sidewalk scales first appeared in 1885. The scales of the next 20 years were big and massive with cast iron bases and tall wooden or cast iron cabinets. They were replaced in the 1920s by sleek porcelain big head "Lollipop" scales. The Depression brought about a change to the smaller personal coin operated scales. All of these scales were placed where they would be seen and used by people. By the 1960s, the coin operated scale business was almost dead. Scales were abandoned on the street, vandalized, and thrown out. Instead of putting your penny into the scale, you went to the department store and bought your own bathroom scale.

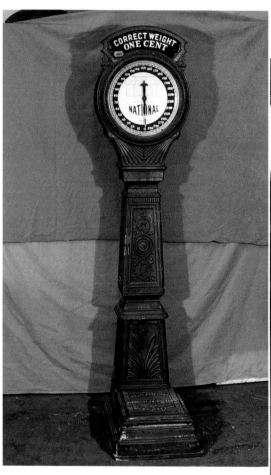

This very ornate cast iron National scale is among the earliest coin operated scales made. It stands 70" tall and has a porcelain face and marquee sign. These scales were very popular in train stations and were almost always painted silver. $2500 - $3000

Eagles are cast into the front of this very ornate U.S.S. Scale Company, cast iron scale. Also cast into the front column is "PROPERTY OF U.S.S. SCALE CO., BOSTON, MASS., U.S.A., NOT TO BE SOLD." $1500 - $1700

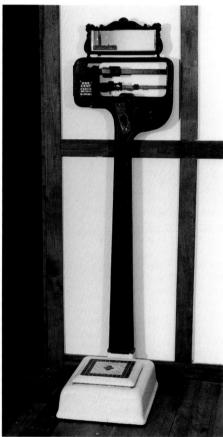

Manufactured by the Caille Company of Detroit, this very accurate beam scale stands over six feet tall. It is made of very ornate cast iron with a porcelain and tile platform. $1200 - $1500

Watling Scale Company made this "Extra Fancy Weighing Scale Style Nine." This is one of the earliest scales Watling made, patented in 1902 and 1903. The dial is etched in gold in the glass face. Inside the casting you can see where the CB for Caille Brothers was removed and replaced with a "W" for Watling. This W is visible on the front of the ornate cast iron column. $3000 - $3500

Shown here is a Watling Manufacturing Company "Profit Sharing, Fortune Telling, Mirror Scale." It has a tall wooden cabinet with a cast iron base. This Watling scale from the 1920s gave you almost everything: your weight, your fortune, and a play field for the coin to drop through with a chance to get it back. This scale is almost identical to style number 16. $1900 - $2200

This is a cast iron National Novelty Company scale. These scales were very popular in train stations. They usually stood outside the small depots year round, which is the reason why they had a fresh coat of silver paint brushed on every year. There were many different marquees made for this scale, with this being the largest. $1500 - $1700

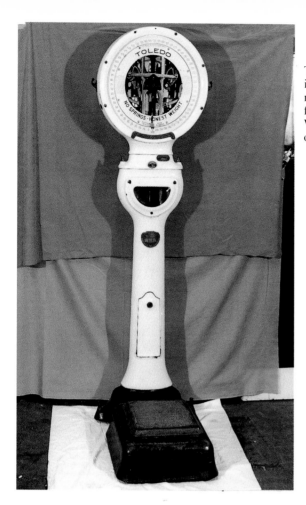

The Toledo model number 8300 T is a large cast iron lollipop scale. The very accurate weighing mechanism is left visible. The Toledo slogan on the front proudly proclaims "NO SPRINGS - HONEST WEIGHT." Toledo also made a free weight version of this scale. $800 - $1200

Watling Manufacturing Company made this style number one, guessing scale. The entire mechanism is nickel plated. All of the lettering on the glass is mirrored, and the inside of the back door is aluminum, polished to a mirror finish. All these special features added together make the head almost glow. There are three coin slots to choose from on the top. If you put your coin into the correct slot, it is returned in the cup below the head. $2200 - $2800

Avery made this English coin operated scale. Made of very heavy porcelain-coated cast iron, it has a capacity of 20 stones and works on a large English penny. $500 - $600

The top of the line Watling Manufacturing Company scale was this "style number 7, Gypsy Fortune Telling Black Cat Scale." The knob is turned to the question you want to ask and one cent gets you the answer and your weight. It is made of porcelain on cast iron with a capacity of 300 pounds. $3600 - $4000

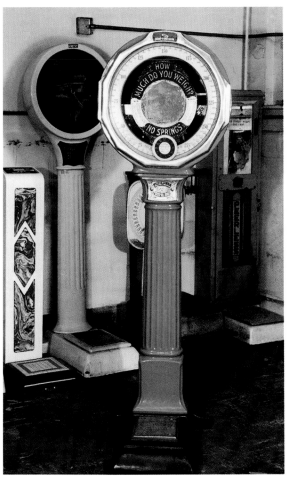

This blue porcelain lollipop scale made by the Watling Manufacturing Company is the "Senator" model. It did not have the gambling feature of the "Ambassador" or the fortune telling capabilities of the "President." It was just a plain weighing scale. $2900 - $3200

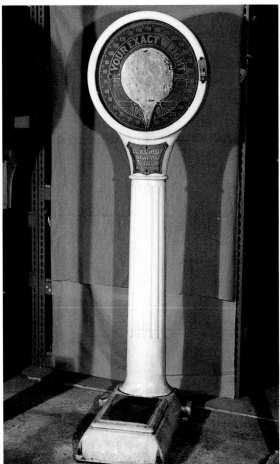

One of the largest lollipop scales is this Mills Accurate. It stands 6'3" tall, with a 22" diameter head. Made in Chicago, in the late 1920s, it has a capacity of 300 pounds and is made of cast iron with a brass face. $700 - $1000

The Watling Scale Company manufactured this 73" tall, mirror front, oak cabinet spring scale. It has a 300 pound capacity and is featured showing off its reflective front. $900 - $1200

This coin-op scale is still taking in coins near Notre Dame in Paris. It was manufactured locally by SAFAA (Société Anonyme Franchise Des Appareils Automatiques). This is a very heavy scale made of cast iron and steel. $900 - $1000

Similarly, this wooden cabinet scale also has a full length mirror front. Here, the user could get his weight and fortune for one cent. It was made by the Watling Scale Company, which is not the same as the Watling Manufacturing Company, and stands over six feet tall. $1500 - $1700

This rare National ticket scale prints your weight on a ticket. Inside this tall narrow scale is an electric motor to run the scale mechanism and print the ticket. $450 - $500

Buy candy and get your weight with this Jennings scale. One coin slot is for weight and the other is marked "sweets." It was made in Chicago and has a capacity of 280 pounds. $1000 - $1200

Shown here are two sizes of Peerless ticket scales. You could drop your coin in the slot and receive a small ticket with a picture of a movie star, your weight, a fortune, the date, and maybe a store ad. Peerless was not a manufacturer of scales; they were an operator. These scales come in a wide variety of colors, with and without chrome trim and mirrors. They are very heavy with many parts. Left: $400 - $450; right: $300 - $400

This is the short version of the
Camco ticket scale. It has the same
mechanism as the large ticket scales.
If you dropped a coin in, you would
receive a movie star ticket with your
weight and fortune. $500 - $550

This version of the Peerless ticket scale has two
mirrors and a tile covered platform. It has a
capacity of 25 to 300 pounds. For one cent, you got
your weight and fortune printed on a ticket with the
portrait of a movie star. $400 - $450

Shown here is a Caille Moderne
scale with a health chart marquee.
This very heavy scale is hard to find
in any condition, especially with the
porcelain marquee. No other scale
has this distinctive tile pattern on
the platform. $450 - $650

This cast iron scale was made by National Automatic Machines Co., Saint Paul, Minnesota. It stands 46" tall. Most coin operated scales, including this one, have separate service and cash box doors in the rear. $350 - $450

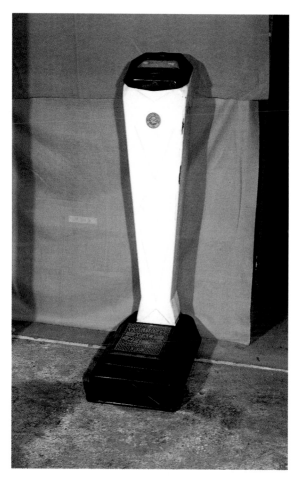

The American Scale Mfg. Co. of Washington, D.C., produced the scale in this picture. This porcelain on cast iron scale is one of the earlier scales made by American. The green circle is a Columbus, Ohio, weights and measures inspection sticker. $250 - $300

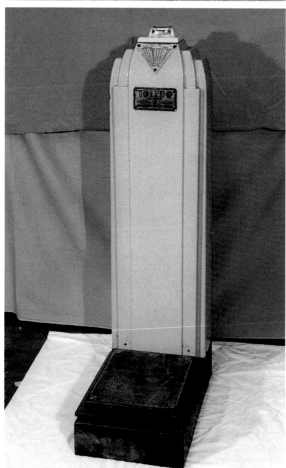

This is the tan version of the Toledo Scale Company's model 8400. It also came with a light inside to illuminate the dial. It is very heavy cast iron and stands 43" tall. $350 - $450

Rockola made this Loboy Personal Weight Scale. This is a rare Rockola that stands a mere 33.5" tall. The reason that it is so short is unknown. Made of cast iron, all known examples are green porcelain. $450 - $550

American Scale Manufacturing Company of Washington, D. C., made this fortune scale, model number 401-M. The M designates the marquee. This model was equipped with six coin slots. If you guessed the lucky slot, your coin was returned. This scale with the marquee is very difficult to find. $300 - $400

Both of these coin operated scales are made by Rockola. The one on the left was made for the Vendinator Corp. of Cleveland, Ohio. The one on the right is a Rockola Loboy Personal Weight Scale. Both are white porcelain over cast iron and stand about 43" tall. These are from the 1930s and were also produced in green. $200 - $350

This small coin operated scale was made by Health-O-Meter in the early 1930s. This battery operated scale is an adapted bathroom scale to which an upright column was attached, allowing the scale to accept coins. It has weight charts for men and women etched in the green glass front plate. Affixed to the scale is a 1934 scale license from the city of Chicago, signed by Mayor Kelly. $200 - $250

This Watling Small Guesser came in several colors. This scale allowed you to guess your weight so your coin would be returned. If you guessed incorrectly, it gave you your correct weight, but kept your coin. The inside of this iron and steel scale is filled with many moving parts. $450 - $650

Shown here is a Watling "Star, Horoscope, Ticket" scale. For one cent and a pull of the handle, you could get your weight and horoscope. This was Watling's top of the line scale in 1940. Very few were made and they were discontinued after the war. $500 - $800

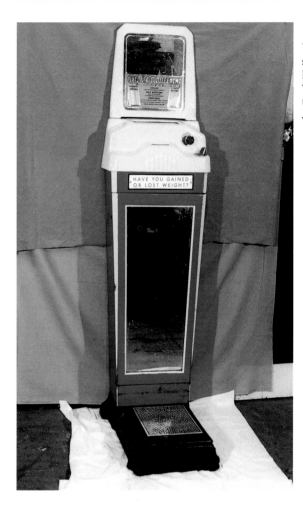

Watling Scale Company made this model 200 Fortune scale from about 1947 to 1952. The user would turn the knob to his choice of over 200 questions for example: "Who will marry me?" After depositing a penny, the user would get his weight and the answer: "A minister or Justice of the Peace." $350 - $450

In the mid 1950s, these Watling Horoscope scales were manufactured by the thousands for use in F. W. Woolworth and other dime store chains. If the scale has the chrome plated hand rails, it most likely came from a Woolworth store. For a nickel, you could get your complete horoscope printed on a scroll, which was rolled up in a plastic sleeve. The scale would then reveal your weight. This was later changed to a dime, then to a quarter. If you did not want the horoscope, you could get your weight for only one cent. $500 - $700

This is a Watling Tom Thumb scale with a marquee designed by a New York operator. The marquee is made of cast iron with a milk glass sign advertising Doctor's Hospitals. $400 - $450

This is a Watling Tom Thumb Junior scale with a charity donation slot. The addition of a donation slot in the center of the front column is the only difference between this model and a non-charity Tom Thumb Junior scale. Most of these cast aluminum scales were made from 1946 to 1972. $350 - $450

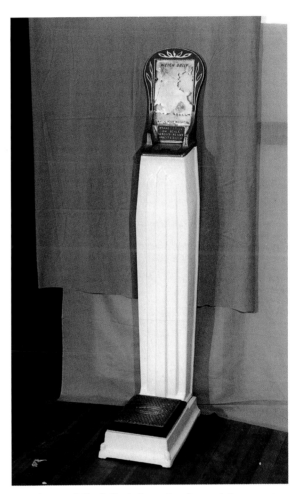

This Pace Mir-O-Scale is made of porcelain on cast iron. This model was so named because of the mirrored marquee and was produced in Chicago in the 1930s. $350 - $500

Bill Watling's Pioneer scale, shown here, was made for a very short period. They were more decorative than the small Watling scales but similar in mechanical design. If the user was able to guess his weight, his penny was returned. This example is missing the marquee. $500 - $600

There are two known examples of this Pace ticket scale. For one cent fate would guide your finger to push the correct button, and you would receive a small card with your fortune printed on it. Your weight was also shown on the scale for the same penny. These scales were made in the 1940s. $1000 - $1200

The Royal 90 pound porcelain on cast iron scale was made in Jersey City, New Jersey. It was available in four colors—ivory, blue, green, or white—and sold for $75 in the 1930s. $300 - $400

This is the old Pace scale "modernized" by the Frantz Company. A foot vibrator was added on some models, and advertising space on the front of the scale was sold for even more income. These scales were also used for charity. The coin slot was enlarged to accept any coin up to a half dollar, and the proceeds went to the charity advertised on the front of the scale. These scales are from the 1940s and 1950s. $275 - $350

O. D. Jennings manufactured this rocket ship advertising scale in the 1930s. This photo shows how the scale looks with no advertising behind the large glass. When in operation, four panels of advertisements illuminate from the inside and are rotated by a motor inside the scale. This heavy cast iron scale is coated with two-tone green porcelain enamel. It has a capacity of 280 pounds and an upper window contains a men's and women's height and weight chart. $2800 - $3100

This is a small cast iron coin operated person scale made by O. D. Jennings & Company, Chicago, U.S.A. It has a capacity of 280 pounds and was also available in other colors. $300 - $400

"RX Scale Manufacturing Company, Buffalo, New York" is the address on the height and weight chart of this scale, and "New Haven, Connecticut" is written on the step plate. Its capacity is 285 pounds. The cast bronze counterweight on the coin mechanism is in the shape of the letters RX. $425 - $475

Hamilton Scale Company of Toledo, Ohio, manufactured these soda bottle scales: Royal Crown Cola, Grapette, and Upper 10. They were made of plastic in the 1950s. Others to look for are Pepsi and Squirt. $3000 - $3500

Here is a Hamilton coin operated person scale with a small marquee. This scale has a cast iron base and a porcelain enameled steel column. It stands 45" tall. It was also available in other colors. $250 - $350

Hamilton Scale Company also made this much sought after coin operated Mr. Peanut scale for use by Planters in front of their Planters Peanut stores. It is believed that less than 100 of these scales were ever produced in the 1950s. Many were broken up so that no one is sure exactly how many are extant today. These scales should have original Bellock keys marked "HS." There should also be matching numbers on the serial number plate of the hat and inside. Beware fakes do exist. $18,000 - $20,000

This Borden's Homo Milk carton advertising scale was manufactured by the Hamilton Scale Company of Toledo, Ohio. This scale was coin operated and still has the coin slot, which was later changed to a free weight. $3000 - $3400

This is a Kirk K-80 high model astrology scale. This is one of the largest coin operated scales made. Mike Munves offered this scale in their 1956 catalog. This photo was taken at a West Coast scrap yard in 1997. Hundreds of scales were saved from the crusher. You can also see an International ticket scale, a Watling lollipop, several Peerless ticket scales, and a few more low model Kirks. $200 - $300

Shown here is a low model Kirk astrology scale. One cent would buy your weight and astrology reading. This scale sold for $75.00 in the Mike Munves catalog during the 1950s. $400 - $600

Guess your correct weight on this ABT guesser and your weight is free. There is a small recess in the bottom of the scale so that your toes don't hit the scale when you stand on its turtle shaped base. $425 - $500

These Pepsi scales were made in Ohio, in the early 1990s. For 25 cents, this digital scale gave you your weight and lucky number. They were made of heavy steel with stainless steel bases to operate flawlessly for many years. It would be difficult for the collector to pry one of these money making scales out of the hands of an operator. $3000 - $3200

This scale was made by C. R. Kirk & Company. Guess your correct weight on this scale and get your money back. At 6" tall, this big sheet metal scale had space for advertising on the front. $300 - $400

108

This is an early version of a Watling lollipop scale. This scale is very difficult to service, because it has no door with which to gain access to the front of the mechanism. It is also a free weight scale, which makes it very rare. $1500 - $1700

Here is a Watling advertising scale. This free weight scale is commonly called a bank scale because most of them were used in banks. The area near the top, shaped like an old TV screen, held a lighted display, usually for a savings account. These scales were made in the early 1960s. $500 - $600

A week before this book was due to the publisher, several of these VEND-R-SCALES were discovered. They had been in storage for almost 30 years in an old basement speak-easy under a corner drug store. To use the scale, you would stand on the platform, put a nickel in the gum machine, and watch the colored liquid go up the thermometer-like tube to indicate your weight. Your height was also indicated on the scale's mirror. These scales were manufactured by the Wico Company in Chicago in the late 1960s. $300 - $400

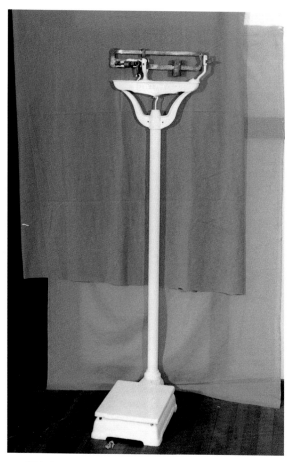

This is a Fairbanks Home Health scale, model number 1263. It measures 58.5" from floor to the tip of the beam. This cast iron scale has a brass beam and a capacity of 300 pounds. $100 - $150

Fairbanks made this doctor's scale with a brass step plate. It is made of wood with an iron base and wheels, and has a 400 pound capacity. $200 - $300

The Fairbanks folding or "Army and Navy" scale is designed for easy transportation or shipboard use. It came in several sizes priced from $58.00 to $72.00 in the 1918 Fairbanks catalog. $250 - $300

This English person scale is from an unknown maker. It still has its original paint with vibrant details visible. It has a capacity of 24 stones and is 40" tall. The base measures 11" wide x 23" deep. $100 - $200

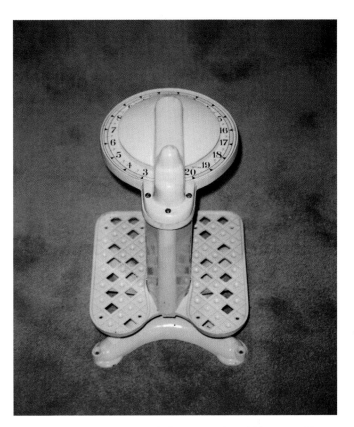

This English bathroom scale has a capacity of 20 stones. You stand on the foot pads of this spring scale and read your weight on the dial. $200 - $300

Avery of Birmingham, England, made this hard to find person scale. Made mostly of wood, it has brass beams and the mechanical parts are of cast iron. The graduations on the beam are calibrated in stones. $325 - $400

The dial on this cast iron bathroom scale is backwards requiring the user to look into the attached mirror to see his weight. It has a capacity of 125 kilos and was made in Europe. $90 - $115

This Universal Bath Scale was made by Landers, Frary & Clark, New Britain, Conn. It is a spring scale made of cast iron with a leather mat on top. It has a capacity of 300 pounds and stands 8.5" tall with a base of 7.25" x 10.5". $30 - $40

Shown here is a Roto Health-O-Meter Dial Scale. This heavy bathroom scale is made of steel and iron with a rubber mat on top. It is equipped with a magnifying lens to help the user see the dial, which has a capacity of 250 pounds. The base measures 10" x 7.5" and the overall height is 8.25". It was made in the U.S.A. $30 - $40

This Counselor scale was made by the Brearley Co. of Rockford, ILL., U.S.A. This bathroom scale is a spring scale made of stamped steel. It has a capacity of 250 pounds and is 6.25" tall and 14" front to back. There is an adjusting screw in the back to zero the scale. $15 - $25

Shown here is a Check Weight Health Scale. This cast iron bathroom scale has a capacity of 250 pounds. The original paint was white, and it has a brass zero adjusting nut on the back. The base measures 9.5" x 11.5", and its overall height is 10". $30 - $40

Shown here is a Hanson Modern Apartment Scale. This 250 pound capacity bathroom scale has a chrome plated platform, which makes it a model no. 735. Model no. 733 has an enamel finish. In the undated Hanson catalog from about 1937, model no. 735 sells for $4.50. It was available in white, green, black, or ivory. The base size is 12.75" x 8", and it is 3.75" tall. $15 - $25

This beam scale was made for weighing a baby. It has an iron cradle which holds a wicker basket in which to lay a baby for weighing. It is unmarked and mounted on a chair sized wooden stand with a sliding poise on a brass beam. $100 - $200

Chatillon made this "Fold-A-Way Baby Scale" which is quite unusual with a tray that can be folded up and removed. It has a capacity of 30 pounds and is 11" tall. To prevent tipping when in use, baby scales usually have larger bases than family scales of a similar type. $35 - $45

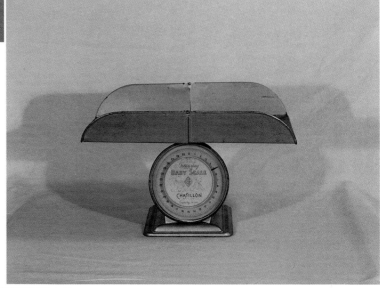

John Chatillon and Sons of New York made this baby scale with a 24" long wicker basket. It is white with gold trim, has a capacity of 25 pounds, and is 15" tall. $45 - $55

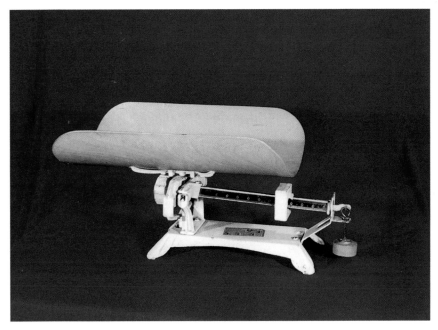

Detecto of Brooklyn, New York, made this cast iron baby scale. It has a bentwood tray, a 30 pound capacity, and is 11" tall. $30 - $40

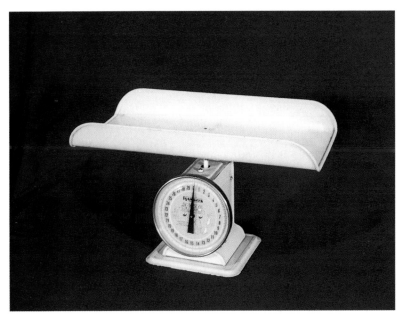

Hanson of Northbrook, Illinois, made this nursery scale with a pink and blue face. It depicts a baby, blocks, and toys on its 30 pound capacity face. This is Hanson's model 3025 and is 12" tall. $15 - $20

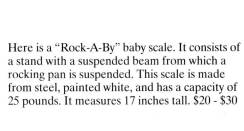

Brearley of Rockford, Illinois, made this baby scale. This thin steel scale is 7" tall and was made in pink and yellow in addition to the blue example shown here. $10 - $15

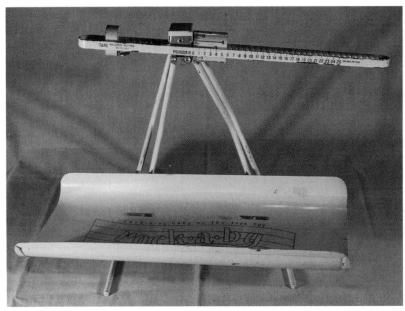

Here is a "Rock-A-By" baby scale. It consists of a stand with a suspended beam from which a rocking pan is suspended. This scale is made from steel, painted white, and has a capacity of 25 pounds. It measures 17 inches tall. $20 - $30

These scales were used mainly in the kitchen to weigh ingredients for favorite recipes. In some European countries today, many cooks still weigh their ingredients. These scales were made from all types of materials such as brass, cast iron, plastic, porcelain, and even tin.

This Krups Ideal scale was made in Germany and is marked "Fabrique Marke D.R.P." This scale rarely exists with the original porcelain pan shown here. It has a capacity of 10 kilograms. $200 - $250

This German kitchen scale has no markings. It is graduated from 0 to 10 kilograms or 0 to 20 pfund. The porcelain tile face is done in tan, light blue-gray, cobalt, and black, and the spring scale is made of cast iron. $90 - $125

This scale is marked "AWR" and has a capacity of 10 kilo or 20 pfund. It is believed to be of German origin and has the old original pan. $85 - $105

George Salter and Company, Ltd. of West Bromwich, England, made this family balance. It has a 14 pound capacity and a 5" diameter brass dial. This ornate cast iron scale has an oblong pan that is 10" x 7". $100 - $125

This Salter's Improved Family scale, class II, has a capacity of 28 pounds. This design is based on Silvester's patent and is rendered in cast iron. The scale pan has a flange, a raised or projecting rim or edge that is friction fit onto the top of the scale. $80 - $100

Shown here is an American Household scale made by ASCo. It has a capacity of 25 pounds by 2 ounces and has a painted tin plate face. Its cast iron body was originally painted blue and yellow. $65 - $85

These two Gilfillan scales are almost the same. They differ in their painted designs, found on the sides and back, and their different brass faces. One of the scales claims it is an "anti-friction scale." Both have capacities of 25 pounds and are 10" tall. $140 - $160

Krups of Germany made this nickel plated brass scale. It has a porcelain face, a 10 kilo capacity, and ball feet. $150 - $175

John Chatillon and Sons made this cast iron spring scale. It is "The New Favorite Family Scale" and has a capacity of 25 pounds. $85 - $100

This Shalers Family Scale dates from 1865. The capacity of this cast iron scale is 24 pounds. The round brass dial face of the scale is shown on the left hand side. $350 - $400

Cast into the heavy iron body of this family scale is "North Bros. Mfg. Co., Phil." and "Pat. 11/6/1877, 6/23/1881." The round brass face is calibrated from 0 to 20 pounds and has "US Scale" and "Pat'd. 1877" printed on it. $100 - $125

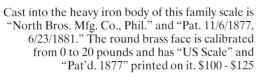

This George Washington scale is made out of stamped steel with a picture of Washington in the center. Unfortunately, Mr. Washington is very difficult to see with the dial hand covering most of his face. This scale is 8" tall and painted green. $30 - $40

The American Family Scale, shown here, was manufactured by American Cutlery Co. of Chicago. It is a relatively common scale that has a capacity of 24 pounds with ounce graduations. $20 - $35

This Accurate scale was made in Western Germany. It is shown with its original pan and has a 25 pound capacity. $35 - $55

Here is a Swedish 10 kilogram scale marked "Jernbolaget." The large pan on top uses the 10 kilogram scale and the small pan on the side uses the gram scale. There is an anchor symbol cast into the front of the scale and enameled on its face. $45 - $55

R. A. Hunter Scale Co. of Cornwells, Pa., made this 25 pound capacity scale from stamped steel. It is painted black and white and has an adjusting screw on the bottom. The Homewate Utility Scale by Royal, Trenton, N.J., looks almost the same as the Hunter scale. The main differences are the 20 pound capacity, the zero adjust screw in the center of the dial, and the color, green and white. Both scales are 6.5" tall with a base size of 5.25" x 8". $15 - $30

This kitchen scale, marked "Justa-Arista," was made in Europe. It has a capacity of 10 kilos and stands about 6.5" tall. $20 - $30

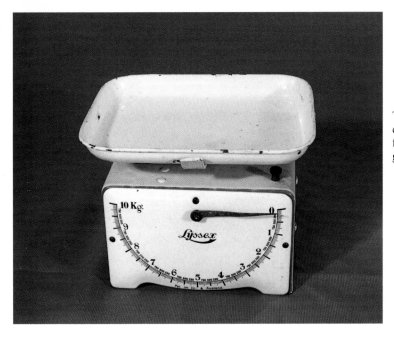

This Lyssex kitchen scale was made in Europe. It has a capacity of 10 kilograms, with a base size of 5" x 7". The face is porcelain and all versions are cream colored with green, blue, or red details on the dial. $30 - $40

The cast iron scale, shown here, has no name or markings on it. Cast into the base is "patent applied for." It is graduated from 0 to 7 pounds on the brass face. It is 10" tall and has a base diameter of 6.5". $140 - $160

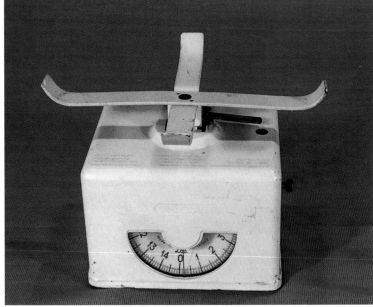

This Inca cast iron kitchen scale was made in Switzerland and has a zero adjusting screw on the side. The base is 5" x 6" and the pan is missing in the photo. $15 - $25

House-Proud is the only name on this scale. The only markings on the face are "made in U.S.A." and "Registered Design." It is made of red plastic with an aluminum cup and has a capacity of 7 pounds. It is 6.5" tall and 7.75" x 3.25" at the base. $30 - $40

Turnbull made this rare cast iron "Novelty Scale" which was patented on July 24, 1877. The arch shaped dial is brass and has graduations from 0 to 12 pounds by 2 ounces. $250 - $400

This Maid of Honor scale is steel and painted white with red trim. It has a capacity of 24 pounds and is 6.75" tall. $20 - $25

Shown here is a Pelouze Domestic scale. This 12 pound capacity scale is made of steel and iron with a brass face. It is painted black with red and white trim. The three patent dates are from 1896 and 1898. It has a zero adjusting screw under the pan. The height is 10.5". $50 - $70

The Aldeco Household scale (*left*) was "made in U.S.A. by Algonquin Tool & Die Co. for Allied Development Corp. Chicago, ILL." This information as well as "Patents Pending" is stamped on the steel bottom of the scale. An identical 25 pound capacity Dazey scale (*right*) was "made by the Dazey Corporation, St. Louis, MO. U.S.A., patents pending." Both are 6.25" high, made out of stamped steel, and painted the same colors, red and white. A chrome version was also produced. The Dazey has three slots near the top. A larger tray might have been attached at these slots, possibly for use as a baby scale. $40 - $70

This scale was made by Gilfillan Scale and Hardware Co. of Chicago. It has a 12 pound capacity and is made of iron and steel with a brass face. It has a zero adjusting screw under the pan and is 8.25" tall. $100 - $120

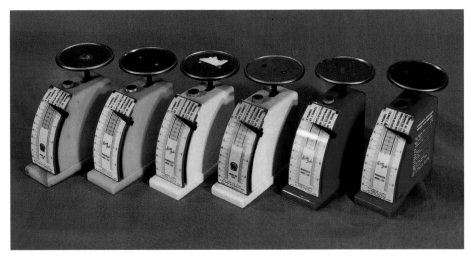

Shown here are several variations of the Hanson Recipe scale, model 1308. The face on these scales is copyright 1948. The early scales were made in Chicago; then production was moved to Northbrook, Illinois. Available in red, white, or yellow plastic with a capacity of 8 pounds. They are 6.5" tall. $15 - $20

This large "candlestick" scale was patented by L. E. Brown in 1878. It has a capacity of 25 pounds and stands about 11.5" tall. These were produced, as shown, in zinc and also in the rarer brass model. $50 - $70

Chicago Scale Co. manufactured this "Little Detective, for family, office, or store." It is 5.5" tall and made of cast iron with a brass beam. The example, shown here, has its original paint and is in good condition. $100 - $125

This English "Registered Family Scale" has a porcelain coated iron pan and a weight plate. Made of cast iron, it has a set of iron weights and a 14 pound capacity. $200 - $250

This German kitchen balance is marked "Herzog" and "5 kilo." Its base is 15.25" long. This scale has a brass beam and sliding poise. $100 - $135

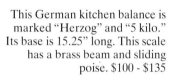

Shown here is a German balance scale with porcelain coated pans. It has a capacity of 5 kilos, and the porcelain body has a windmill and sailboat Delft style design. $150 - $250

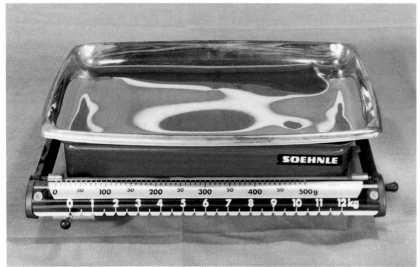

Soehnle, a German company started in 1868, made this 12 kilogram scale. Most of it is made of steel, with a few plastic and aluminum parts. The Soehnle logo on the front is in a style that was first introduced in 1963. $20 - $30

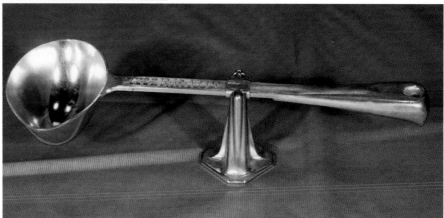

This ladle scale was made in Italy. It is made of cast aluminum with a capacity of 7 ounces or 200 grams. It is 12.75" long. $30 - $45

Shown here is an aluminum and stainless steel ladle scale made in Austria. It has a capacity of 8 ounces or 100 grams. The bowl is also graduated in cups with a capacity of one cup. $25 - $30

Shown here is a Roman Meal Diet scale with a capacity of 230 grams by 5 grams. Lines inside the scoop carry volume measures of one-forth, one-half, three-quarters, and one cup. Made of plastic, it is meant to be held between the thumb and forefinger, allowing measured ingredients to be easily emptied into a bowl or cooking dish. $10 - $20

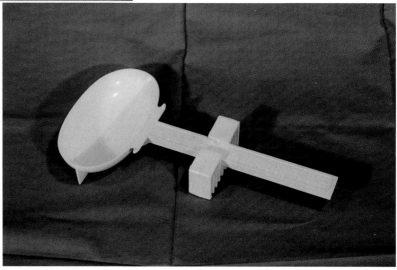

These tiny scales are well liked by many collectors. Some were made as exact replicas of large existing scales and were used as samples by salesmen, while others were made especially for children to play with.

Shown here is a German tin-plate scale marked "Jouet." It has a 150 gram capacity. $40 - $50

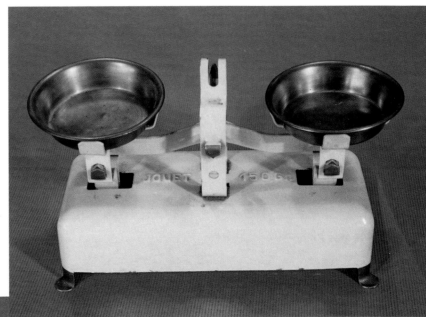

This is a small, tin, toy balance with brass pans. It has no name or markings, but was probably made in Germany. The base is 2.5" x 5.5". $35 - $40

Shown here is "The Little Shopkeeper's Toy Scales." The base is only 4" long on this British made toy scale. It is stamped from light weight tin and has no markings other than those on the box. $10 - $15

This small scale and desk set on a wood base has no name or markings. The wood base measures 6" x 4.5". The pencil and weights are missing. It is most likely of German origin. $75 - $90

The "Play Store Computing Scale" was made by Kingsbury of Keene, New Hampshire. It is made from stamped steel and painted red and white. It has a zero adjusting lever in the back and is graduated in ounces with a capacity of 16 ounces. This is one of the few toy scales that works accurately. It is 6.5" tall. $35 - $45

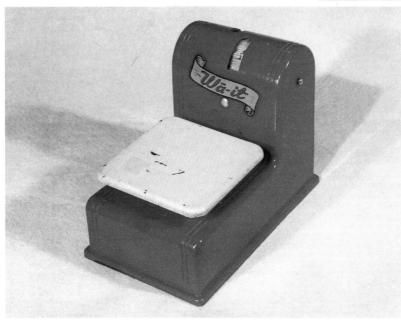

Shown here is a Wa-it scale made by the Al-Vern Corp., of Jerome, Mich. It indicates patents pending and pats applied for. It also has "not to be used for commercial weighing" printed on it. This scale is made of red plastic with the tray made of stamped steel, painted white. It has a capacity of 5 pounds and is 6.5" tall with a 5.25" x 8.5" base. $15 - $25

This American toy scale is made from painted tin and has nice graphics of children using the scale. This example is missing the dial. The scale has a battery operated light inside. $30 - $40

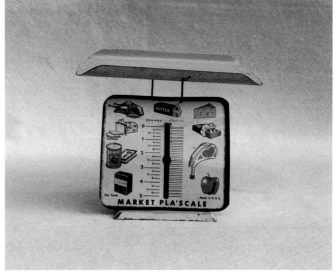

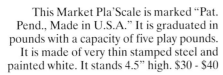

This Market Pla'Scale is marked "Pat. Pend., Made in U.S.A." It is graduated in pounds with a capacity of five play pounds. It is made of very thin stamped steel and painted white. It stands 4.5" high. $30 - $40

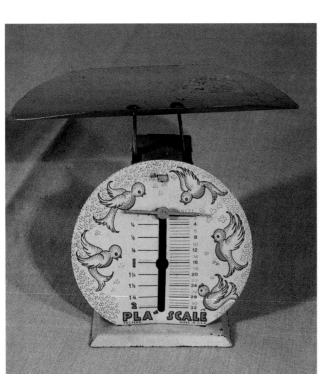

Here is a "Pla' Scale, Pat. Pend., made in U.S.A." It is graduated in pounds and ounces with a capacity of 2 pounds. It is made of stamped steel, painted pink, and is 4.5" in height. This scale is very similar to the Market Pla'Scale. $30 - $40

This Gilbert toy scale was included in a 1960s chemistry set. It is made of tin and steel with a 1.5" x 6" base. $15 - $20

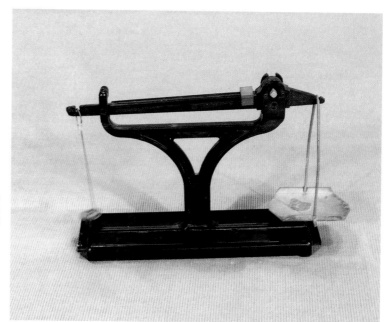

The Kiddie scale is made from stamped steel and painted green. It is a very well made scale for use as a toy. It has a zero adjusting screw on top and an overall height of 5.75". $25 - $30

Here are two cast metal toy scales with no names or markings on them. $20 - $25

Here is a ceramic scale bank marked "Lego" and "Japan." On its front is "Honest weight, Save Here, Honest Savings to go a-weigh with." When the user drops a coin into the slot marked "Coins," the scale's pointer moves. It stands 5.75" tall. $10 - $20

This cast iron scale with a cast aluminum pan is not very old. It is very crudely made and almost works. One similarity between this scale and the real ones is that the small weights are often missing. $5 - $12

This Berkel computing scale is a salesman's sample or toy. It is very well made and about 4" tall. The immobile weight and goods pans are steel with the scale itself being made of porcelain enameled steel. It is believed to be of English origin. $100 - $110

These English made lead toy scales are about 3" tall. They appear to be copies of early Salter coin operated scales. $30 - $45

This American plastic "Bank Balance" has no markings to indicate its manufacturer. It is designed to look like a pillared bank with doors and a clock. The beam is steel, while the weight hangers are copper. $5 - $15

This chapter covers a wide range of scales, which don't exactly fit into any other category. Some of these scales create their own category because they are so very rare. Within this category, there are textile testing scales, paper scales, sport related scales, large truck scales, steelyards, and many other types that you may not know exist.

Shown here is an unmarked fishing contest scale or fishing club scale. It was strictly used for the purpose of weighing fish in fishing contests. $250 - $300

This unmarked pocket knife has a 12 pound capacity scale incorporated into its design. $100 - $125

This Fishweigh fish scale, made in Cary, North Carolina, is the perfect tool for the well equipped tackle box. You simply hang the fish from the correct notch located on top of the knife blade; then slide the handle out until the fish balances. The weight of the fish is indicated on the exposed knife blade. $250 - $350

These Langley Fisherman's De-Liar fish scales are quite small and include a tape measure. The larger size, model 228, has a capacity of 28 pounds, while the smaller, model 208, has an 8 pound capacity. There are many variations of these scales made of both plastic and metal and under many different names in the U.S.A., Japan, England, and other countries. $8 - $15

The Redding powder and bullet scale has a capacity of 325 grains. Made in Cortland, New York, the original price on the box is $14.00. The base is cast iron. $35 - $50

Carter Mfg. Co. of Rockford, Illinois, made this cast aluminum fish scale. It has a capacity of 24 pounds. About 100 of these scales were made in the 1940s and all were sold locally. $40 - $60

The Fisherman's Pal is a 30 pound capacity fish scale. It has a 10" ruler for honest fishermen, a 20" ruler for dishonest fishermen, and a stainless steel knife. This floating tool is made in Japan. $20 - $30

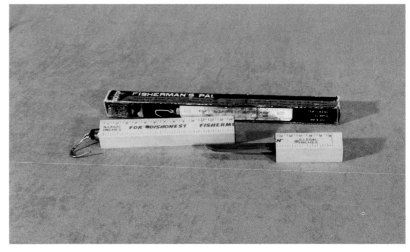
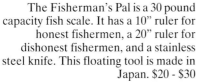

Shown here is a Prorythmic golf club scale made by Kenneth Smith Golf Club Co., of Kansas City, Mo. It is graduated from A to G, 0 to 625, and 0 to 22. It is made of aluminum with a stainless steel poise. This scale's total length is 25.5". $110 - $150

Fairbanks manufactured this powder scale which is missing its pan. It has a capacity of 250 grains on the nickel plated beam. $90 - 110

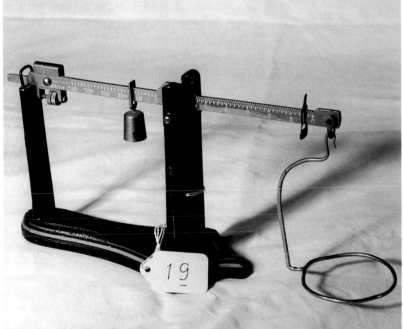

Shown here is an inexpensive unmarked scale. It fits into its own plastic box which also serves as a stand. $5.00 - $15.00

This English, Sky Weighs luggage scale was made in both red and brown. Both hooks fold neatly around the leather covered scale to enable it to fit in its own leather carrying case. The scale has a capacity of 65 pounds or 35 kilos, and was used to weigh luggage for air travel. $15 - $30

This scale is marked "Patented Prorythmic Golf and Tennis Balance" and "Kenneth Smith Golf Club Co. Kansas City, MO." It was used to determine the head weight of a golf club and the balance of a tennis racket. $180 - $210

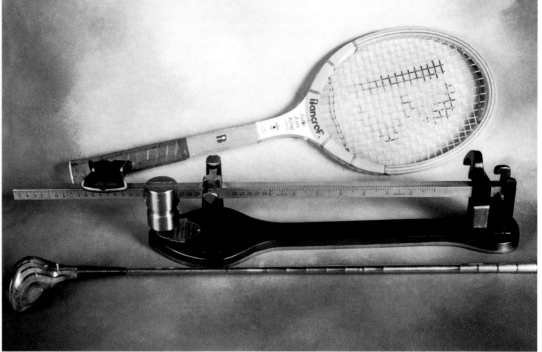

This Food portion scale was probably made by Hanson Scale Co., for Edward Don & Company, Chicago. The purpose of this scale is to establish menu prices for restaurants. By following the cost-per-pound charts on the face and using the formula above them, the restaurateur can establish minimum menu prices. The entire face of this one pound scale rotates so that zero can be set at any position for a tare. $25 - $35

This "Ice Cream Scale" was probably made by Hanson Scale Co., for R. K. Merritt & Associates, 1100 West 8th, Azusa, Calif. This is a model 1509M with a capacity of 5 pounds, graduated in scoops (ounces) from 0 to 80. The total height of this scale is 9.25". $35 - $50

Lyman Metal Products of Norwalk, Conn., made this Martini Scale for mixing martinis in three different ratios: 5:1, 10:1, and 25:1. Instructions are printed on the beam for mixing the gin and dry vermouth. The two aluminum jiggers can be lifted from the beam. The wood base measures 5.25" x 3". At least one other larger size scale was manufactured. Another one of these scales has been seen with the name "Loyal Gift Products Inc., N.Y.C." $30 - $45

E. J. Hoadley of Hartford, Connecticut, made this 2 pound capacity scale. It is made of cast iron and sheet brass with a spring mechanism and an adjustment screw. The more common version of this scale has a round plate and a hammered brass dish. $90 - $125

This 5" high scale made by the Pelouze Co. in Evanston, Ill, was used for ice cream. It has a capacity of two pounds, is graduated by 0.5 ounces, and also shows the percentage of overrun with 8 ounces equal to 100%. $20 - $30

Landers, Frary, and Clark made this Universal Ice Balance. It bears a patent date of 1917 and has two scales. $35 - $50

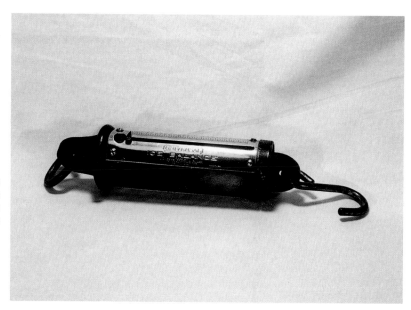

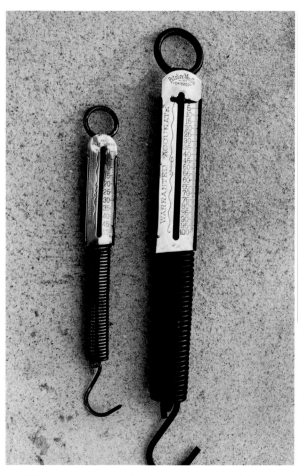

Shown here is a grouping of small, brass, and nickel plate faced, spring balances. Their faces range from 3" to 6" in length and they have capacities of one-half pound to fifty pounds. They were made by Chatillon, Salter, and other makers. $8 - $25

These interesting spring balances were made by Pelouze Mfg. Co. of Chicago. They are made of iron and brass and are unlike other hanging scales of this type in that the springs are exposed. The small one has a capacity of 50 pounds, while the large one weighs items up to 100 pounds. $35 - $50

This Eastman studio scale was made by Eastman Kodak Co., Rochester, N.Y., U.S.A. The base measures 4" x 9". The metal parts on these scales are often rusty or corroded due to harsh photographic chemicals. $50 - $60

Fairbanks made this very early "model # 13" platform scale. It is made of cast iron with a brass beam marked "Pittsburgh Novelty Works." This rare scale was originally painted black with red trim. The beam is graduated from 0 to 7 pounds. $600 - $800

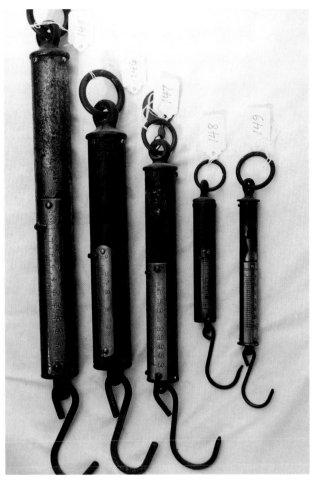

Here is a grouping of iron, tubular spring scales. Each have a rectangular shaped vertical brass plate graduated from 0 to 25, 0 to 50, or 0 to 100 pounds, and range from 0.75" to 1.5" in diameter. $15 - $35

This rare spring scale has no markings as to the maker. Once painted black, it bears a brass, crescent shaped dial with a capacity of 200 pounds. $175 - $200

Chatillon made this brass faced spring balance. It has a capacity of 10 pounds by sixteenths of an ounce and measures 4.5" wide x 8" long. $110 - $125

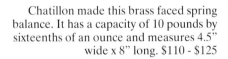

This is a cast iron "Fairbanks Domestic." It has a ceramic goods pan and an octagonal weight plate. This 6" tall scale is painted black with red, copper, and brass colored details. $90 - $125

This is a Russian wood and brass bismar. Marked 1812, it has sustained many years of hard use and has some old repairs. The pivot for this bismar is missing. $75 - $150

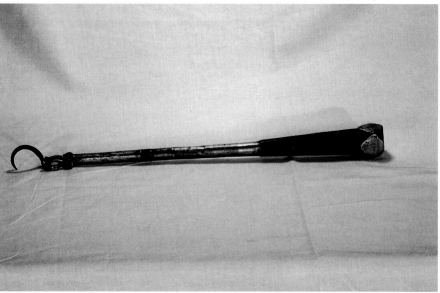

The two mancurs shown here are most likely of European origin since both weigh in kilograms. The giant mancur (*left*) has a single capacity of 110 kilograms and is 14.5" long. The smaller mancur, 9" long (*right*), is a dual action scale with two graduations of 18 kilos and 110 kilos, one on each side. Each have their own hook, are rare with an angular shape, and made of iron with brass dials. $175 - $250

Some antique dealers incorrectly refer to this as a Chinese opium scale. The actual use for this scale is to weigh coins, gold, and silver. The ivory beam and brass pan fit in the violin shaped wood box. Many of these are old, but some are still manufactured for tourists. $45 - $65

Shown here is a V-spring balance. This is an iron, French, 18th or 19th century scale. It is marked "C. Hamo" and has a 15 kilogram capacity. The wooden box makes a decorative frame for this hard-to-find scale. $75 - $125

This brass steelyard was made by J. S. Trowbridge Co., 16 Elm St., Boston. It is 18" long and has a capacity of 55 pounds. $125 - $140

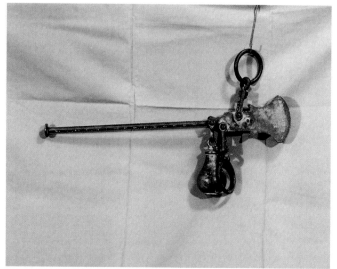

Wm. B. Preston, No. 14 Portland St., Boston, made this brass steelyard in about 1890. The weight, hooks, and rings are of iron and it has a capacity of 60 pounds. $135 - $150

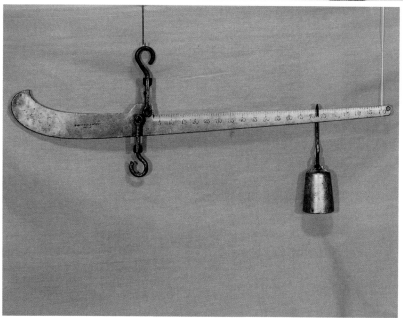

Henry Troemner of Philadelphia made this 16 pound capacity butter scale. It is made of porcelain enameled iron and has a brass beam. The weights for this ball scale are made of tobin bronze. The porcelain goods plate has a diameter of 10". $150 - $200

The Eagle Scale Works of Philadelphia manufactured this brass steelyard. It has a capacity of 200 pounds on one side and 100 pounds on the other with the four pound brass weight. The total length is 32". $150 - $200

This iron steelyard is hand forged with a capacity of 12 pounds on one side and 55 pounds on the other. Most hand forged steelyards, including this one, are without a name or markings. $25 - $40

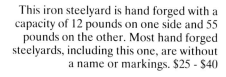

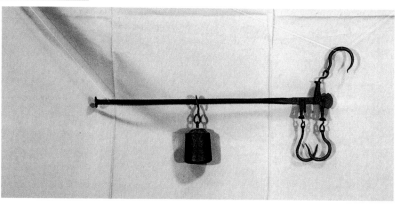

The Torsion Balance Co. of New York made this cream testing scale. It has a copper finish and bears patent dates of 1885, 1889, and 1891. $60 - $110

This is a Fairbanks cream testing scale. The four pins on the plate held a tall bottle. $95 - $115

This Jones of Binghampton cheese factory scale was used in a factory or creamery to weigh milk or cream. By using the five beams, milk from five different producers could be weighed with a separate beam for each one. The milk would not have to be emptied with each new load. $400 - $500

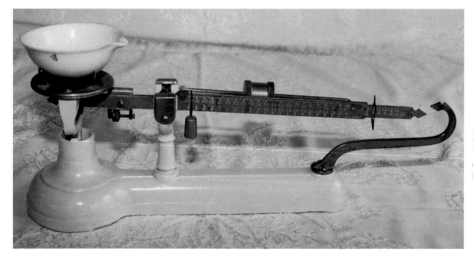

Shown here is a Fairbanks butter testing scale. This scale is used for determining the amount of water in butter or fat. $125 - $150

Renfrew Scale Company of Renfrew, Ontario, made this grain scale. It is made of maple, and cast and sheet iron. It has a capacity of 2000 pounds shown on its brass beams. Extra weights hang on a bar on the leg. It has a patent date of 17, October 1911. $700 - $800

The "Dollydale" scoop scale, by the Robson Corporation of New York City, has a capacity of 5 pounds. It is made of aluminum and has a wooden handle. The indicated weight of the goods being weighed changes as the scoop is held at different angles. $30 - $40

The Penn Scale Mfg. Co., Inc. made this poultry scale. It came equipped with a cone shaped, open ended pan for weighing live poultry. $30 - $45

Fairbanks Morse & Co. made this 1500 pound slaughter house track scale. These are the beams used to weigh dressed meat as it moved around the slaughter house on over head tracks. All the large iron levers would have been suspended above the track system. With a capacity of 1500 pounds, this would have been a small scale. $40 - $50

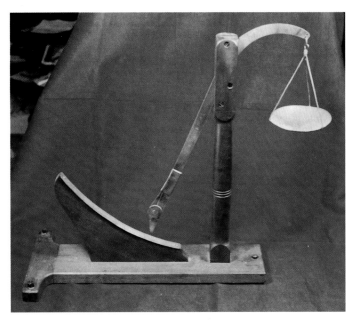

The name "A. G. Spindler" is hand stamped on the pointer arm of this hand made scale. It has a capacity of 50 grams. There are three leveling screws in the 16" x 5" wooden base. The purpose of the scale is unknown. $35 - $45

The Smith Scale Co. of Columbus, Ohio, U.S.A., made this "Exact Weight" over/under scale. This type of scale would be used to spot check pre-packaged goods, like sugar or flour, sold by the pound. $80 - $100

This tea sampling scale has no name or markings. It sits on a wooden base with a drawer for the weights. $175 - $200

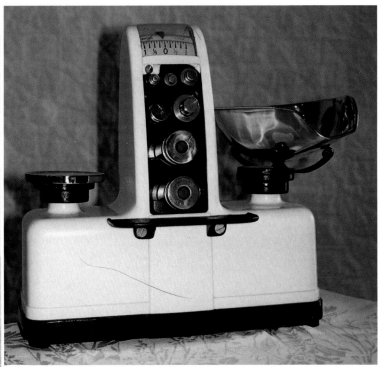

This Detecto over/under scale has a complete set of weights and a 3 pound capacity. $60 - $80

S. C. W. S. Ltd. of Edinburgh, Scotland, made this coal or potato scale. It has a 28 pound capacity and stands 38" tall. $200 - $300

"National Scale Company, Weighing and Counting Machine, Chicopee Falls, Mass., U.S.A." is cast into a plaque on the front of this scale used for counting parts. Small parts are placed in the pan at the top, parts to be counted are placed on the platform, and the total amount is computed on the brass beam. $375 - $425

J. Abeletor and Son made this English inverted Roberval Imperial scale. It has a capacity of 56 pounds. $250 - $300

Stark Brothers Ribbon Corp. distributed this Stark's Inventory scale to take inventory of boxed ribbon. The scale is screwed onto a wooden box with a flip over lid. The instruction card, which is used to set the scale on zero, is the same weight as an empty ribbon box. This scale calculates the length of the ribbon remaining in the box. The box is 6.25" tall with the lid closed. $120 - $145

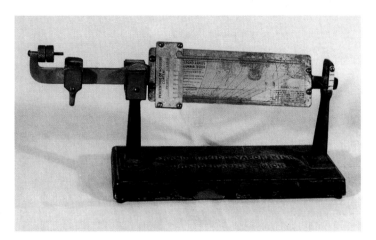

This is a Buffalo Scale Company moisture percentage scale for lumber. This scale was used to calculate the percentage of moisture in a lumber sample. Complete instructions are on the brass beam. The brass pan is missing in this photo. $110 - $130

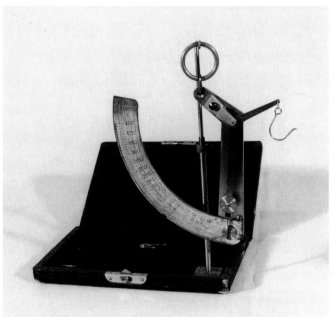

This yarn scale was made by Amthor Testing Instrument Co., of Brooklyn, N.Y. It was used to test the weight of yarn samples and is graduated from 100 to 1000 feet per pound. The entire scale comes apart to fit into its case. $90 - $110

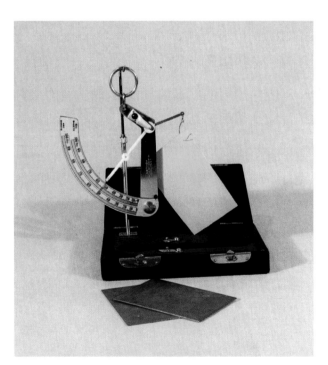

This scale made by Amthor Testing Instrument Co. Brooklyn, N.Y., was used to weigh paper samples. The two brass plates are used to cut out the sample paper to size. The entire scale comes apart to fit in the case. $80 - $100

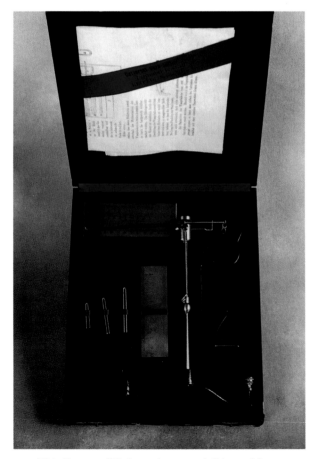

This German "Universal yarn scale" is used for testing yarn. The scale and stand fit into the box along side all of the attachments. It was made in the late 1800s. $350 - $400

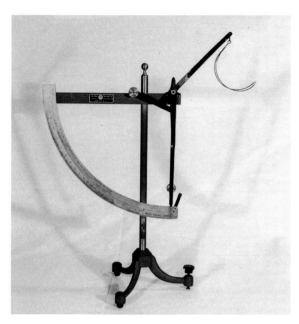

This yarn scale was manufactured by Alfred Suter Co., a textile instrument company in New York City. It measures grains per yard in a 5 yard sample from 50 to 70 grains. $80 - $110

Shown here is a Chatillon board scale. This scale is used to weigh pasteboard to make up bundles of 50 pounds. $110 - $130

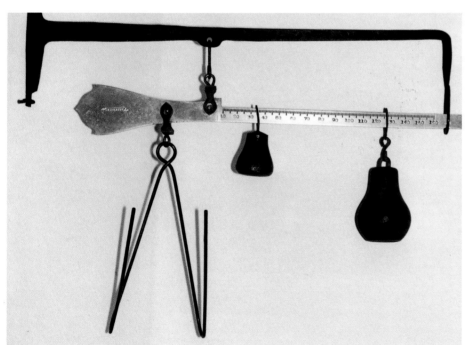

Fairbanks made this paper scale. One side of this brass steelyard is calibrated from 0 to 40 pounds and the other is calibrated from 10 to 160 pounds. It was made to weigh a ream of paper based on the weight of a few sheets. This one is shown with its original suspension bracket. $160 - $180

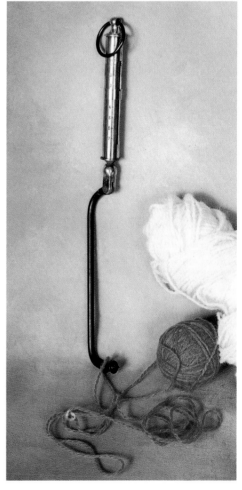

This is a yarn testing scale. This scale has a capacity of 7 kilos. You can see the brass indicator ring about half way down the brass cylinder. $90 - $100

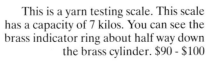

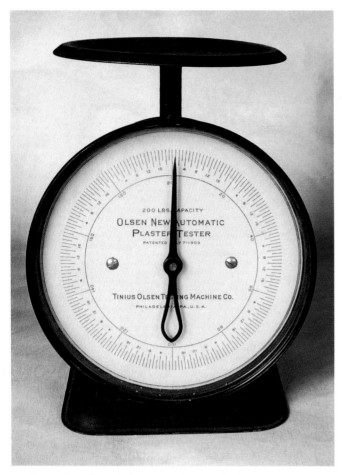

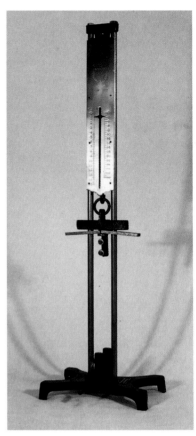

This device is not a weighing scale, but has a Chatillon scale built into it. It is used to test the breaking strength of string or wire. The wire is hooked to the scale and the crank. When the crank is turned, the scale will stop at the point where wire breaks. The capacity of this tension scale is 15 kilos or 30 pounds. It stands 30" tall. $40 - $50

Shown here is an "Olsen New Automatic Plaster Tester." It was made by Tinus Olsen Testing Machine Co., Philadelphia, Pa., U.S.A. $55 - $65

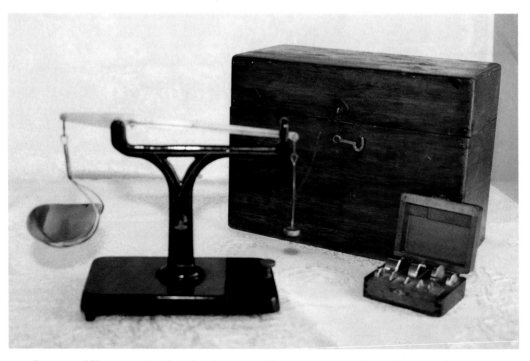

Brown and Sharpe made this estimating scale with a complete set of cased weights. Once disassembled, the scale and cased weights can be stored in the fitted wooden case. $150 - $175

This glass blower's scale, with a round brass plate, was manufactured by Fairbanks. There are two brass beams, each with a brass poise and a locking screw on the bottom. $150 - $170

Chattilon made this heavy duty 1000 pound capacity scale. Made of very heavy cast iron, it has a brass dial that is 10" in diameter. $50 - $70

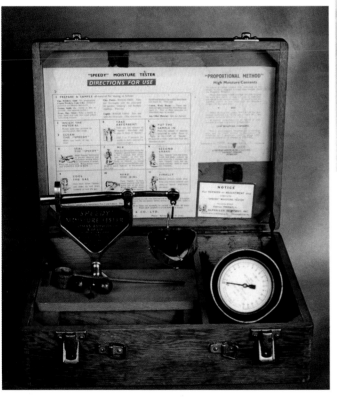

This Speedy Moisture Tester was made in England by Thomas Ashworth and Co. Ltd. The complete scale and all its accessories fit into a wooden box, which also houses instructions for the scale's use. $150 - $170

The purpose and age of this Toledo scale is unknown; although, it is quite similar in design to a Toledo moisture testing scale. It has a capacity of 1000 and is very well made. $25 - $50

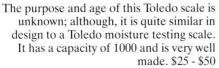

Humboldt Manufacturing Company of Germany made this balance that is mounted on a pail. It was used to weigh items in a liquid for specific gravity calculations. It stands 26.5" tall. $35 - $45

Degrave of England made this interesting portable inspector's scale. Shown here in its specially fitted wooden box, it is extremely accurate in order to test weights. $550 - $650

This Fairbanks counting scale is used for counting small parts, screws, bolts, nails, and other hardware items. It calculates ratios of 100:1, 50:1, 25:1, or 20:1. One item is placed in the small pan and a set amount of the same part will balance the scale. $600 - $650

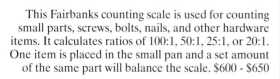

This chapter includes a variety of ideas for expanding your scale collection beyond scales. Incorporated here are signs and other forms of scale company advertising, catalogs, real photo postcards, fobs, desk furniture, and pins.

Resembling a scale, this is actually a radio made to look like an existent scale. Concealed in the base is an AM transistor radio. It was made in Japan and is plastic and metal. The base measures 6.75" x 3.5". It operates on one nine volt battery. $100 - $150

Here is a nickel plated brass miniature scale pan commemorating Fairbanks Scales 100th anniversary, 1830-1930. This was used as a paperweight, candy dish, desk piece, or, most often, an ash tray. If it was used as an ash tray, it would probably have burn marks in the bottom of the pan. The base measures 3" x 3", and the pan measures 6" long. $70 - $80

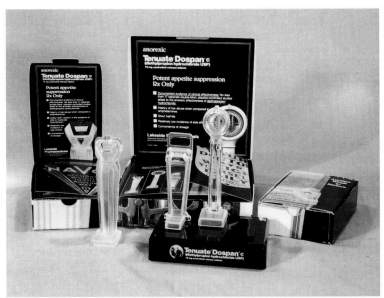

Lakeside Pharmaceuticals sent this display of three different plastic coin operated scale models to doctors as part of a weight reducing drug advertisement and contest. The three winners of the contest each received one of the actual antique scales pictured. They were the Peerless Aristocrat Deluxe, Navco, and the Watling 400 Deluxe. Each model was sent out in a separate mailing, with the display stand for the models included in the first box received. The tallest model is 6". These were mailed out in 1988-89. $80 - $100

This is a 50th Anniversary paper weight with "Dayton Scale Division, The Hobart Mfg. Co., Troy Ohio., 1891-1941" printed on the top. This cast iron piece measures 3.75" in diameter. The top part is stamped out of very thin brass. $40 - $50

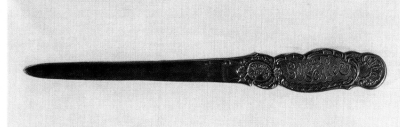

This solid brass letter opener advertises "Buffalo Scale Co." The reverse side of the handle has a cherub on it. $40 - $50

The Fairbanks Company of Philadelphia gave away this 12" ruler. It folds together at the center to fit into the leather pocket case. $20 - $25

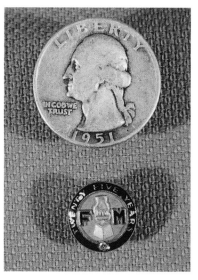

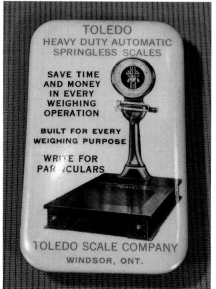

This advertising piece given away by the Toledo Scale Company, Windsor, Ont., is a honing stone. It measures 1.75" x 2.75". Pictured on the front is a large Toledo platform scale. $30 - $40

These key chains were given away by the Watling Scale Co. of Chicago. These are made out of plastic covered paper with an insert to record the owner's name and address. They measure 1.75" long. The scale shown is the Tom Thumb Junior. Top: $25 - $30; bottom: $8 - $10

Shown here are three pocket watch fobs. The Fairbanks Morse one, on the right, is made of a silver colored metal and bears the Fairbanks hand held weight symbol. It is inscribed "Fairbanks Morse Quality, 100%" and is designed to look like an old wax letter seal. The two fobs on the left are from Dayton Moneyweight Scale Co. They are both made of gold colored metal and have a computing barrel scale depicted on their faces. The lower fob is framed in blue enamel with a white enamel scale plate and the Dayton Moneyweight scale sign in red enamel. This fob also depicts, on its obverse, an eagle holding a set of computing scale beams with the words "Dayton Moneyweight Scales" written in an open circle around the eagle and scale. Dayton: $100 - $125; Fairbanks: $70 - $80

Fairbanks Morse Co. gave this 10 karat gold and diamond pin to their employees after twenty-five years of service. It has a blue enameled edge and bears the hand held weight symbol, F. M. It measures 0.5" in diameter. $50 - $60

152

This is a Fairbanks Morse Co. badge from the Beloit (Wisconsin) Works. It bears the hand held weight company symbol and appears to be nickel plated over brass. It measures 1.25" across. $20 - $30

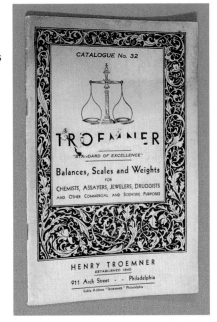

This Troemner catalogue, *Balances, Scales and Weights for Chemists, Assayers, Jewelers, Druggists and Other Commercial and Scientific Purposes*, is number 32. It is dated 1937, and contains 64 pages. It is 6" x 9" and contains illustrations of over 60 scales. $60 - $70

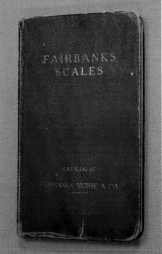

This *Fairbanks Scales,* catalog 67, is dated 1918. It is 4.5" x 7" and contains 198 pages. This catalog is marked "Fairbanks Morse & Co." and has scale illustrations on most pages. $75 - $100

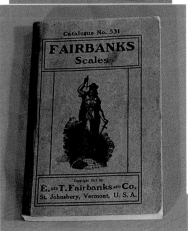

Here is a *Fairbanks Scales*, Catalogue No. 531. It is 4.5" x 7", paperback, with 250 pages, and over 200 illustrations of scales, weights, parts, and accessories. $75 - $100

These undated *Buffalo Scales Miniature Catalogs* are believed to date from the 1920s and 1930s. The catalog on the left contains 64 pages, the center one contains 28 pages, and the right hand one contains 58 pages. All of these catalogs are profusely illustrated. $25 - $40

This Henry Troemner *Scales and Weights*, Catalogue No. 1926, has 152 pages. It measures 6" x 9", has over 100 illustrated scales, and is dated 1926. $75 - $100

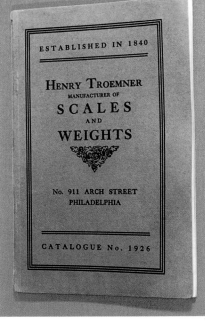

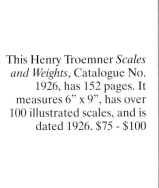

W. T. Avery of England may have used these photo etched printing dies for the scale pictured in one of their catalogs, flyers, or other advertisements. The etching was done in copper and then mounted on wood. The average size of these dyes is 2" x 3" and they are almost 1" high. $3 - $10

The wooden sign on the left is an advertisement for Forsyth & Co. and measures 37" x 19". This scale company was in business, according to city directories, only from 1863 to 1870. The Buffalo Scale Co. wooden sign measures 28" x 16". These wooden signs generally predate porcelain signs. $150 - $250

This wooden "Fairbanks Scales" sign is identical on front and back. It may have been on a large scale or used for advertising. It measures 13" x 42". $150 - $250

The "Jones Scale Works, Binghampton NY" sign is cast iron and was said to have been originally on the bridge approaching the factory. It measures 16.5" x 7". The stencil "From Fairbanks Morse & Co. St. Johnsbury, VT. Made in U.S.A." is made of thin brass. It was used to stencil the return label on shipping crates. "The Richardson Scale Co. Clifton, N. J. U.S.A." sign is made of blue and white porcelain on steel and measures 32" x 2.5". The wooden "Fairbanks Morse & Co. Sales - Service" sign is painted dark green with gold letters. It measures 36" x 10". Jones: $175 - $225; Fairbanks: $40 - $50; Richardson: $90 - $110; Fairbanks sales: $140 - $160

The Toledo sign, at the top center, is cobalt blue and white porcelain over steel. It measures 28" x 12". The "Winslow Scale" sign is made from heavy tin sheet metal. It is painted green with gold letters and measures 48" x 10". The "Cardinal Scales" sign measures 40" x 8" and is porcelain enameled in white with red letters. The two shields were made by Peerless Scale Co. for use on their coin operated Aristocrat and Aristocrat Deluxe scales. The shield with the police man on it (*right*) is very desirable but has been reproduced. Toledo: $140 - $175; Winslow: $40 - $60; Cardinal: $70 - $80; Blue shield: $80 - $100; Police shield: $300 - $350

Shown here is a "Buffalo Scales" porcelain on steel sign. It is white with blue letters and measures 30" x 12.5". $120 - $150

This sign is believed to have been on display in a store where only Toledo scales were used. It reads "Our Weighing Service is rendered by Toledo Scales No Springs-Honest Weight." It is white with black letters and is 17.5" x 11". $120 - $150

This "Fairbanks Scales" is blue and white porcelain over steel. It measures 60" x 8". $100 - $125

Shown here is a "Fairbanks-Morse Sales and Service" electric back-lit sign. It is made of frosted glass with a decal affixed to it. The decal bears the hand held weight company symbol rendered in white, black, yellow, and green. It has a housing made of thin sheet metal which encloses an electric light fixture. The entire sign is hung by a chain which is affixed to the upper sides of the housing. This sign is 14" in diameter. $150 - $250

The location of the butcher shop in this real photo postcard is unknown. A Toledo computing scale sits on the counter next to a tray of fresh asparagus. $30 - $35

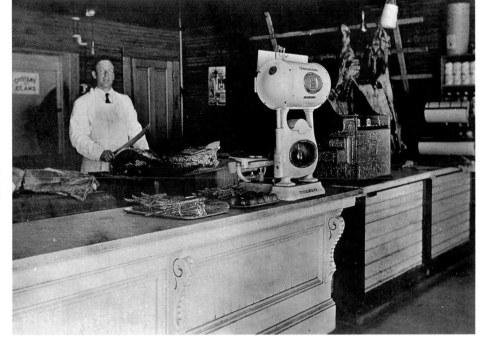

This is a real photo postcard of the Charles T. Saul Pharmacy, 2758 Wyoming Street, in an unknown city. Standing by the front door is a coin operated scale. It was manufactured by either the National Novelty Company or the Mills Novelty Company. $45 - $50

A pair of Hamilton Mr. Peanut scales stand at the front door of the Planters Peanut store. This color postcard is from the Chattanooga Tennessee store. $5 - $8

Shown here is a real photo postcard from California. A Toledo coin operated scale stands in front of the souvenir shop in the redwood forest. $8 - $12

This real photo postcard is a store front view of the Moneyweight Scale Company. An assortment of Dayton Moneyweight scales and several scale weights are displayed in the front window. The reflection in the front window shows the photographer with his camera. $35 - $45

Building Your Collection

How do you start your very own collection of scales? The first thing to do is to decide what type of scale to collect. Important things to take into consideration are the amount of space you have to devote to display or storage of scales, the amount of time you have for the quest for interesting scales, and the amount of money you can spend for scales. In the event that you have a few empty rooms and a big bucket of money, you could easily buy well over 500 different scales in the first year. That's only ten scales per week. In order to do that you would have to buy almost any scale, in any condition, and at any price. Most collectors will want to be more selective than this. Thus, though there are numerous types of scales readily available in varying conditions, better scales require more time and money to be found.

Many collectors will only collect one type of scale or scales from one area. Some for example choose to limit their collections to egg scales, letter scales, coin scales, brass faced spring balances, or Fairbanks scales, while others collect only scales from the state they live in.

Hopefully, you already have a collection started and know in which direction you would like to go. If not, you will know soon!

Shown here and on the following two pages are several different scale collections. These will give you ideas of how some veteran collectors display their own collections. There are many other options for displaying a collection. The type and size of scales you collect will have some bearing on your method of display. Many brass faced scales, for example, are meant to be hung and display well on a background of rustic barn board or dark paneling. Elegant Tiffany postals lend themselves to a more formal setting than family scales. Every different type of scale requires its own specialized size, depth, and height of shelves. Most collections start small and grow one scale at a time. Therefore, there is the need to leave space for a collection to grow. Once a collection gets to 'the point of no return,' many collectors decide to build or create a special "scale room" in the house. Scales can also be displayed in various ways in many different settings throughout the house. Bathroom, coin operated, or doctor's scales display well and are useful in a bathroom. A confectioner's or shop scale is well displayed in a kitchen. Whether one collects a scale for its aesthetic beauty or because it is mechanically interesting is the collector's choice.

Resources

The following are not only what we used in writing this book, but are very valuable to all collectors of scales. They are listed here for you to explore other avenues of interest in scale collecting.

Crawforth, Michael A. *Handbook of Old Weighing Instruments.* Chicago, Illinois: International Society of Antique Scale Collectors, January, 1984.

Bietry, Dr. L. *Dictionary of Weighing Terms: A practical guide to the Terminology of Weighing.* Mettler Instruments AG, 1883.

Bueschel, Richard M. *Big Head Lollipop Scales.* Fountain Valley, California: "Coin-Op Classics Magazine," 1994.

Bueschel, Richard M. *Collector's Guide to Vintage Coin Machines: With Price Guide.* Atglen, Pennsylvania: Schiffer Publishing Ltd., 1995.

International Society of Antique Scale Collectors. *Equilibrium* plus other publications and catalog reprints. 300 West Adams St., Suite 821, Chicago, IL 60606

Jewell, Brian. *Veteran Scales and Balances.* Turnbridge Wells, Kent, England: Midas Books, 1978.

Yale, Allen Rice, Jr. *Ingenious And Enterprising Mechanics: A Case Study of Industrialization in Rural Vermont, 1815 - 1900.* Ann Arbor, Michigan: UMI Dissertation Services, 1995.